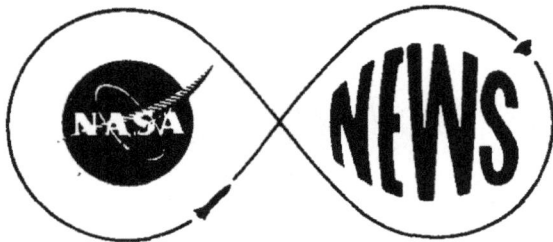

NATIONAL AERONAUTICS AND
SPACE ADMINISTRATION
Washington, D. C. 20546
202-755-8370

RELEASE NO: 72-220K

FOR RELEASE: Sunday
November 26, 1972

PROJECT: APOLLO 17
(To be launched no
earlier than Dec. 6)

I0030529

contents

-more-

November 14, 1972

TABLES AND ILLUSTRATIONS

Cover: Apollo 17 commander Eugene A. Cernan is holding the lower corner of the American flag during the mission's first EVA, December 12, 1972. Photograph by Harrison J. "Jack" Schmitt.

Image Credit: NASA

Published by Books Express Publishing
Copyright © Books Express, 2012
ISBN 978-1-78039-866-2

Books Express publications are available from all good retail and online booksellers. For publishing proposals and direct ordering please contact us at: info@books-express.com

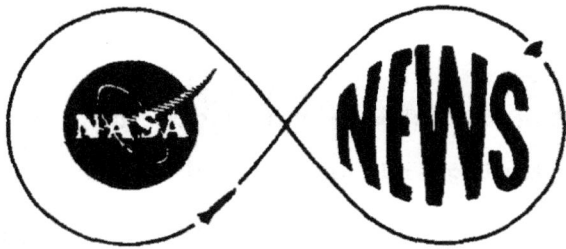

NASA NEWS

NATIONAL AERONAUTICS AND SPACE ADMINISTRATION
Washington, D. C. 20546
PHONE: 202/755-8370

David Garrett
(Phone: 202/755-3114)

RELEASE NO: 72-220

FOR RELEASE: Sunday,
November 26, 1972

APOLLO 17 LAUNCH DECEMBER 6

The night launch of Apollo 17 on December 6 will be visible to people on a large portion of the eastern seaborad as the final United States manned lunar landing mission gets underway.

Apollo 17 will be just one and a third days short of the US spaceflight duration record of 14 days set in 1965 by Gemini VII, and will be the sixth and final Moon landing in the Apollo program. Two of the three-man Apollo 17 crew will set up the fifth in a network of automatic scientific stations during their three-day stay at the Taurus-Littrow landing site.

-more-

In addition to erecting the scientific data relay station, Apollo 17 has the objectives of exploring and sampling the materials and surface features at the combination highland and lowland landing site and to conduct several inflight experiments and photographic tasks.

The Taurus-Littrow landing site is named for the Taurus mountains and Littrow crater located in a mountainous region southeast of the Serenitatis basin. Dominant features of the landing site are three rounded hills, or "massifs" surrounding the relatively flat target point and a range of what lunar geologists describe as sculptured hills.

Apollo 17 will be manned by Eugene A. Cernan, commander, Ronald E. Evans, command module pilot, and Harrison H. Schmitt, lunar module pilot. Cernan previously flew in space aboard Gemini 9 and Apollo 10, while Apollo 17 will be the first flight into space for Evans and Schmitt. Civilian astronaut Schmitt is also a professional geologist. Cernan holds the rank of Captain and Evans is a Commander in the US Navy.

During their 75 hours on the lunar surface, Cernan and Schmitt will conduct three seven-hour periods of exploration, sample collecting and emplacing the Apollo Lunar Surface Experiment Package (ALSEP). Four of the five Apollo 17 ALSEP experiments have never been flown before.

ALSEP, powered by a nuclear generator, will be deployed
and set into operation during the first extravehicular activity
(EVA) period, while the second and third EVAs will be devoted
mainly to geological exploration and sample collection.

The crew's mobility on the surface at Taurus-Littrow
again will be enhanced by the electric-powered Lunar Roving
Vehicle (LRV). Attached to a mount on the front of the LRV
will be a color television camera which can be aimed and
focussed remotely from the Mission Control Center.

Cameras operated by Cernan and Schmitt will further
record the characteristics of the landing site to aid in postflight geological analysis.

Data on the composition, density and constituents of
the lunar atmosphere, a temperature profile of the lunar surface
along the command module ground track and a geologic cross-
section to a depth of 1.3 kilometers (.8 miles) will be gathered
by instruments in the service module Scientific Instrument
Module (SIM). Evans will operate the SIM bay experiments and
mapping cameras while Cernan and Schmitt are on the lunar surface.
During transearth coast, he will leave the spacecraft to recover
film cassettes from the mapping cameras and the lunar sounder.

Apollo 17 will be launched from Kennedy Space Center Launch Complex 39 at 9:53 pm EST December 6. Lunar surface touchdown by the lunar moduel will be at 2:55 pm EST December 11, with return to lunar orbit scheduled at 5:56 pm EST December 14. After jettisoning the lunar module ascent stage to impact on the Moon, the crew will use the service propulsion system engine to leave lunar orbit for the return to Earth. Trans-earth injection will be at 6:32 pm EST on December 16. Command module splashdown in the Pacific, southeast of Samoa, will be at 2:24 pm EST December 19. There the spacecraft and crew will be recovered by the USS Ticonderoga.

Communications call signs to be used during Apollo 17 are America for the command module and Challenger for the lunar module. During docked operations and after lunar module jettison, the call sign will be simply "Apollo 17."

Apollo 17 backup crewmen are US Navy Captain John W. Young, commander; USAF Lieutenant Colonel Stuart A. Roosa, command module pilot; and USAF Colonel Charles M. Duke. All three have prior spaceflight experience: Young on Gemini 3 and 10, and Apollo 10 and 16; Roosa on Apollo 14; and Duke on Apollo 16.

Summary timeline of major Apollo 17 events:

Event	December Date	EST
Launch	6	9:53 pm
Translunar Injection	7	1:12 am
TV-Docking & LM extraction	7	2:05 am
Lunar Orbit Insertion	10	2:48 pm
Descent Orbit Insertion #1	10	7:06 pm
Descent Orbit Insertion #2	11	1:53 pm
Lunar Landing	11	2:54 pm
Start EVA 1 (7 hours)	11	6:33 pm
TV Camera on	11	7:48 pm
Start EVA 2 (7 hours)	12	5:03 pm
TV Camera on	12	5:31 pm
Start EVA 3 (7 hours)	13	4:33 pm
TV Camera on	13	4:58 pm
Lunar Liftoff (TV on)	14	5:56 pm
TV-LM & CSM Rendezvous	14	7:31 pm
TV Docking	14	7:54 pm
Transearth Injection	16	6:31 pm
TV- View of Moon	16	6:46 pm
Transearth Coast EVA (TV-1 hr)	17	3:18 pm
TV - Press Conference	18	6:00 pm
Splashdown	19	2:24 pm

(End of general release; background information follows.)

A COMPARISON OF LUNAR SCIENCE BEFORE AND AFTER APOLLO

The astronomical observations of the Moon prior to
Apollo give us a very detailed picture of the surface of
this planet. However, even the most sensitive telescopes
were unable to furnish the variety of scientific data that
is necessary to the understanding of the history and evolution of the planet. In particular, it was necessary to know
something about the chemistry and something about the internal state or condition of the planet before we could do much
more than speculate about the origin and past history of the
Moon. The most important scientific observations concerning
the Moon that existed prior to the direct exploration of the
Moon by either manned or unmanned spacecraft are as follows:

1) The mean density of the moon is 3.34 gm/cc. When
this number is compared to the density of other planets (this
comparison involves a substantial correction for the effects
of pressure in planets as large as the Earth and Venus), we
see that the density of the Moon is less than that of any of
other terrestrial planets. If we accept the hypothesis that
stony meteorites are samples of the asteroids, we also observe
that the Moon is lower in density than the parent bodies of
many meteorites. This single fact has been an enigma to any-
one attempting to infer a chemical composition for the Moon.
One thing can be clearly concluded from this fact -- that is,
that the Moon has less metallic iron than the Earth. The
difference between the lunar density and that of chondritic
meteorites is particularly puzzling because these objects
have compositions that are similar to those of the Sun once
one removes those elements which form gaseous compounds at
modest temperatures (hydrogen, helium, nitrogen, carbon,
neon, and the other rare gases).

2) The second major characteristic of the Moon goes
back to Galileo, who observed that the Earth-facing side of
the Moon consisted of mountainous regions that he designated
terra, and smoother, physiographically lower regions which
he designated mare by analogy with the terrestrial oceans
and continents. The albedo or reflectivity of these two
regions is markedly different -- the mare regions being
very dark when compared to the terra regions. Astronomical
studies added a great deal of detail to Galileo's discovery,
including some rather fine features such as the rilles which
were just barely resolved by good telescopes. However, the
cause of this fundamental physiographic difference was not
well understood before the era of Apollo. The explanation
of the relatively smooth mare basins ranged from the conclusion that they were very extensive lava fields to the hypothesis
that they were, in fact, dust bowls -- that is, extensive dust
deposits. There were even some scientists who seriously suggested
that they were filled by a type of sedimentary rock that was
deposited at a very early stage in lunar history when the Moon
had an atmosphere.

-more-

3)   The origin of the circular depressions or craters, which are the most common physiographic feature of the lunar surface, was the basis of continual scientific controversy. Two types of explanations were offered -- first, that they were volcanic features similar to terrestrial calderas or volcanic collapse features; secondly, that they were produced by projectiles impacting on the lunar surface in the way that meteorites had occasionally been observed to fall on Earth. In fairness to the proponents of the various theories, it should be recognized that no one ever claimed that all craters were either meteoritic or volcanic. Those scientists tending to favor the volcanic origin emphasized that large numbers -- including some craters much larger than any terrestrial caldera -- were volcanic in origin. Others favoring the impact origin also admitted that a few atypical craters such as Davy Rille and the dark halo craters of Alphonsus may, indeed, be evidence of minor volcanism on the lunar surface.

4)   In parallel with the role of volcanism on the lunar surface, there were two schools of thought on the thermal history of the Moon. The first of these held that the Moon was a relatively inactive body which may have undergone some chemical differentiation which, in any event, took place very early in lunar history. The second expected that the Moon was similar to the Earth with a long and continuous record of volcanism and chemical differentiation. Some adherents to this school fully expected that some volcanism may have persisted to the most recent geologic epochs; that is, as recently as 10 million years ago.

5)   The chemistry of the lunar surface was a total unknown before Surveyor V. Nevertheless, there were a number of definite suggestions -- for example, it was at one time suggested that carbonaceous chondrites were derived from the dark mare regions of the Moon. Others suggested that a type of meteorite known as eucrites was representative of the lunar surface. Still others suggested that a very silica-rich glass found in mysterious terrestial objects called tektites must represent parts of the lunar surface. One could not even be sure that these hypotheses were all inconsistent with each other. At this point in time, we will never know the extent to which the Surveyor analyses may have affected our understanding of the Moon. The data returned from these analyses were of surprisingly high quality. They were, however, so quickly superseded by the analyses of the returned samples that there was never sufficient time for them to be completely integrated into scientific thinking on the Moon.

6) Several other results obtained by unmanned spacecraft
helped set the stage for Apollo. They are the discovery of
the mascons, which require a remarkably rigid or strong lunar
shallow interior.-- the determination (by Explorer 35) that
the Moon had a very weak, perhaps nonexistent, magnetic field;
and finally, the observation (by both Russian and American
spacecraft) that the lunar backside was very different from
the frontside in that dark mare regions were essentially
absent from the backside of the Moon.

As we anticipate the sixth manned landing on the lunar
surface, we are infinitely richer in facts concerning the Moon.
Many of the facts and observations have already been tentative-
ly assembled into theories and models which are leading us to
a genuine understanding of the Moon's history. In other cases,
it is proving extremely difficult to come up with an explanation
that accounts for all of these facts in a self-consistent way.
The major areas of understanding which have come out of the unmanned exploration and five manned landings are briefly out-
lined here:

1) We now have a rather definite and reliable time scale
for the sequence of events in lunar history. In particular,
it has been established with some confidence that the filling
of the mare basins largely took place between 3.1 and 3.8
billion years ago. Since these surfaces represent the major
physiographic features on the lunar surface, we can immediately
infer that the bulk of lunar history recorded on the surface
of the Moon (that is, the time of formation of more than 90
percent of the craters) took place before 4 billion years ago.
This is quite different from the terrestrial situation where
most of the Earth's ocean basins are younger than 300 million
years, and rocks older than 3 billion years make up an almost
insignificant proportion of the surface of the Earth. One of
the major objectives frequently stated by groups of scientists
involved in planning the Apollo science activities was to
find rocks that might date back to the formation of the Moon,
itself. Underlying this objective was the hope that one might
find a primitive or predifferentiation sample of the planet.
Up until now, this objective has eluded us in the sense that
none of the samples returned to date is unmodified by younger
events in lunar history. There is strong circumstantial evidence (for example, the apparent age of some soils) that rocks
dating back to 4.5 or 4.6 billion years must exist on the
lunar surface. However, it now appears that the intense
bombardment of the lunar surface by projectiles that range in
size up to tens of kilometers in diameter was rather effective
in resetting most of the clocks used to determine the absolute
ages of rocks. The widespread occurence of highland material
with an apparent age of 3.8 - 4.1 billion years is today
associated by some scientists with the formation of the Imbrium
basin, which is thought to be produced by the collision of a
50 kilometer projectile with the lunar surface.

We should recognize, however, that more detailed studies
of some of the returned samples may change the interpretation
of these ages. In other words, we cannot be nearly as sure
that we know the age of formation of the Imbrium basin as we
are of the time of crystallization of the mare volcanic rocks.

2) The relative importance of volcanic and impact-
produced features on the lunar surface is today rather well
established. There is almost unanimous agreement that the
dark mare regions are, indeed, underlain by extensive lava
flows. This is shown both by the rocks returned from the
Apollo 11, 12, and 15 sites and by the high resolution photographs which give us very convincing pictures of features
comparable to terrestrial lava flows. On the other hand,
almost all craters appear to be caused by impacting projectiles.
The occurrence of volcanic rocks in the terra regions is an
open question. Preliminary interpretations of the Apollo 16
samples suggest that volcanic activity in the highland region
may be highly restricted or virtually nonexistent.

3) A major objective underlying many of the Apollo experiments was the investigation of the lunar interior. Most
of our information concerning the interior of the Earth derives
from a knowledge of the way in which the velocity of acoustic
waves varies with depth. The study of terrestrial earthquakes
has provided a detailed picture of these variations within the
Earth. The Apollo 11 seismograph indicated that when compared
to the Earth the Moon is seismically very quiet. This result
is, of course, consistent with the conclusion that volcanism
and other types of tectonic activity have been rare or absent
from the lunar scene for the last 2-3 billion years. It was
a disappointment in that it indicated that information regarding the interior would be sparse. However, the use of SIVB
impacts and the very fortunate impact of a large meteorite on
May 13, 1972, have today shown a remarkable structure for the
upper 150 kilometers of the Moon. In particular, we have
learned from lunar seismology that the Moon has a crust more
than 60 kilometers thick. More precisely, one should say
that there is a seismic discontinuity where the velocity of
sound increases suddenly from 7 kilometers per second to
8 kilometers per second at about 65 kilometers depth. The
precise origin of this discontinuity is still a subject for
debate.

The most commonly held explanation suggests that it is due to the chemical differentiation of the upper part of the Moon -- in particular, that an extensive, partial melting of this region produced low-density liquids which arose to cover the lunar surface, leaving a high-density residue that accounts for the high velocity material below 65 kilometers. A minority opinion holds that the velocity contrast at 65 kilometers is due to a pressure-induced phase change.

4) We now have a much more detailed understanding of the Moon's present magnetic field. It is clearly not negligible as was thought prior to the Apollo missions. The magnetometers emplaced on the lunar surface reveal a surprisingly strong, but variable, field. Both the direction and the intensity of the magnetic field vary. This heterogeneity in the local field is, of course, smoothed out when one moves away from the lunar surface. So the field seen by an orbiting spacecraft is much lower than that recorded on the surface. We have also determined that the mare lava flows crystallized in a magnetic field which was much stronger than that of the present Moon. This raises the very interesting possibility that during its early history, the Moon was either embedded in a relatively strong inter-planetary magnetic field or had a magnetic field of its own which has since disappeared. Either possibility presents very serious problems in the sense that we are forced to make assump-tions which are not entirely consistent with what some scientists hypothesize we know about the Sun or the early history of the Moon.

5) The fluctuation in the magnetic fields measured at the lunar surface is a function of the flux of incoming charged particles (solar wind) and the internal electrical conductivity of the Moon. Careful study of these fluctuations shows that the Moon has a relatively low conductivity. To a first order, the conductivity of most silicates is a function of temperature and chemical composition. It is particularly sensitive to the abundance of ferrous and ferric iron. At present, one cannot completely sort out these two parameters. If the interior of the Moon has a "normal" iron concentration, the conductivities appear to place upper limits of 1200-1500°C on the temperature of the deep interior.

6) The heat escaping by conduction from the interior of a planet depends on the amount of heat produced by the decay of radioactive elements, the thermal conductivity of the deep interior, and the initial temperature of the deep interior. Using rather imprecise models and making reasonable assumptions concerning the abundance of the radioactive elements potassium, uranium, and thorium, it was expected that the energy flux from the interior of the Moon would be substantially lower than that for the Earth simply because the Moon is much smaller.

The first measurement of this quantity at the Apollo 15 site indicates that this is not the case. If this measurement is characteristic of the whole Moon, the only plausible explanation that has been put forth to date requires: First, that the Moon is richer in the radioactive elements uranium and thorium than the Earth; and secondly, that these elements are strongly concentrated into the upper parts of the Moon. When combined with the observations on the volcanic history of the Moon and the present-day internal temperatures, the energy flux leads to two current pictures of lunar evolution. The first assumes that the variation in radioactivity with depth is a primary characteristic of the planet; that is, the planet was chemically layered during its formation. In this case, the initial temperature of the lunar interior below 500 kilometers was relatively low, and the deep interior of the Moon gradually became hotter, perhaps reaching the melting point during the last billion years. Volcanism can be entirely accounted for by early melting in the outer 400 kilometers of the Moon which were formed at a higher temperature than the central core of the Moon. The alternative model of thermal evolution assumes that the Moon was chemically homogeneous when it formed and underwent extensive chemical differentiation to bring radio-activity to the surface very shortly after its formation -- in other words, we begin with a molten Moon. Each of these models has some problems. The objectives are, in a sense, esthetic. Some scientists object to the hypothesis that the Moon was initially heterogeneous on the grounds that such structure requires special assumptions regarding the processes that formed planets. Others object to the idea of a molten Moon that conveniently differentiates to bring the radioactivity to the surface as an equally arbitrary idea.

7) The most extensive and diverse data obtained on the lunar surface are those concerned with the chemistry and mineralogy of the surface materials. The study of samples from the six Apollo sites and two Luna sites reveals a number of chemical characteristics that are apparently moonwide. There is, nevertheless, some hesitancy to generalize from these relatively minute samples to the whole lunar surface. Fortunately, two experiments carried out in lunar orbit provided excellent data regarding the regional distribution of various rock types. The x-ray fluorescence experiment very convincingly defined the prime difference between the chemistry of the mare and highland regions. It showed that the highland regions are unusually rich in aluminum -- much richer, in fact, than most terrestrial continents. This observation, along with the ubiquitous occurence of fragments which show an apparent enrichment in the mineral plagioclase, leads to the strong hypothesis that the regolith and soil of highland regions is underlain by a "crust" similar to the terrestrial rock designated anorthosite.

-more-

The x-ray fluorescence results show that mare regions have aluminum concentrations 2-3 times lower than those of the terra or highland regions, along with magnesium concentrations that are 1 1/2 - 2 times greater than those of the terra regions. These differences are totally consistent with the chemistry of the returned samples. When combined with data from the returned samples, these observations provide an excellent explanation of the morphological and albedo differences. We have, for example, determined that all mare basalts are unusually rich in iron and sometimes rich in titanium. The high iron concentration of the mare vis-a-vis the low concentration of the highlands is the basic explanation of the albedo differences since both glass and mineral substances rich in iron and titanium are usually very dark. A second experiment carried from lunar orbit shows that the region north and south of the crater Copernicus is remarkable rich in radioactive elements. A band going north from the Fra Mauro site to a region west of the Apollo 15 site contains soil that must have 20 times more uranium and thorium than most of the mare or terra in other parts of the Moon. The existence of a rock rich in these elements was also inferred from samples from the Apollo 12, 14 and 15 sites. The uneven distribution of this rock -- commonly disignated KREEP basalt -- is a major enigma in the early evolution of the Moon. The time of formation of rocks with this chemistry is not well determined. There is, however, strong circumstantial evidence that some of the uranium-rich KREEP basalts were originally formed between 4.3 and 4.4 billion years ago. Both the samples and orbital geochemical experiments indicate that the three most common rocks in the lunar surface are plagioclase - or aluminum-rich anorthosites; uranium, thorium-rich "KREEP" basaltic rocks; and iron-rich mare basalts. With the exception of the mare basalts, we do not have well documented, unambiguous theories or models that explain the chemical or mineralogical characteristics of these rocks. Nevertheless, the differences between the lunar rocks and terrestrial rocks are so marked that we can conclude that the Moon must be chemically different from the Earth. The Moon appears to be much richer in elements that form refractory compounds at temperatures of 1600-1800°K. Thermodynamic considerations show that calcium, aluminum, and titanium silicates are the most refractory compounds that exist in a solar dust cloud. Many scientists are now coming to the conclusion that the chemistry of the lunar surface is telling us that some separation of solid material and gas in this dust cloud took place at temperatures in excess of 1600°K. The Moon is also strongly depleted in elements that are volatile at high temperatures. This is, of course, consistent with the enrichment in refractory elements.

-more-

None of the three theories regarding the origin of the Moon -- that is, separation from the Earth, capture from a circumsolar orbit, or formation from a dust cloud surrounding the Earth -- can be absolutely ruled out from the present data. The chemical difference between the Earth and the Moon must, however, be explained if the Moon was torn out of the Earth. The depletion in volatiles and enrichment in refractories place a constraint on this theory that will be very difficult to account for.

APOLLO 17 MISSION OBJECTIVES

The final mission in the Apollo lunar exploration program will gather information on yet another type of geological formation and add to the network of automatic scientific stations. The Taurus-Littrow landing site offers a combination of mountainous highlands and valley lowlands from which to sample surface materials. The Apollo 17 Lunar Surface Experiment Package (ALSEP) has four experiments never before flown, and will become the fifth in the lunar surface scientific station network. Data continues to be relayed to Earth from ALSEPs at the Apollo 12, 14, 15 and 16 landing sites.

The three basic objectives of Apollo 17 are to explore and sample the materials and surface features at Taurus-Littrow, to set up and activate experiments on the lunar surface for long-term relay of data, and to conduct inflight experiments and photographic tasks.

The scientific instrument module (SIM) bay in the service module is the heart of the inflight experiments effort on Apollo 17. The SIM Bay contains three experiments never flown before in addition to high-resolution and mapping cameras for photographing and measuring properties of the lunar surface and the environment around the Moon.

While in lunar orbit, command module pilot Evans will have the responsibility for operating the inflight experiments during the time his crewmates are on the lunar surface. During the homeward coast after transearth injection, Evans will perform an in-flight EVA hand-over-hand back to the SIM Bay to retrieve film cassettes from the SIM Bay and pass them back into the cabin for return to Earth.

The range of exploration and geological investigations made by Cernan and Schmitt at Taurus-Littrow again will be extended by the electric-powered lunar roving vehicle. Cernan and Schmitt will conduct three seven-hour EVAs.

Apollo 17 will spend an additional two days in lunar orbit after the landing crew has returned from the surface. The period will be spent in conducting orbital science experiments and expanding the fund of high-resolution photography of the Moon's surface.

SITE SCIENCE RATIONALE

	APOLLO 11	APOLLO 12	APOLLO 14	APOLLO 15	APOLLO 16	APOLLO 17
TYPE	MARE	MARE	HILLY UPLAND	MOUNTAIN FRONT/ RILLE/MARE	HIGHLAND HILLS AND PLAINS	HIGHLAND MASSIFS AND DARK MANTLE
PROCESS	BASIN FILLING	BASIN FILLING	EJECTA BLANKET FORMATION	• MOUNTAIN SCARP • BASIN FILLING • RILLE FORMATION	• VOLCANIC CONSTRUCTION • HIGHLAND BASIN FILLING	• MASSIF UPLIFT • LOW LAND FILLING • VOLCANIC MANTLE
MATERIAL	BASALTIC LAVA	BASALTIC LAVA	DEEP-SEATED CRUSTAL MATERIAL	• DEEPER-SEATED CRUSTAL MATERIAL • BASALTIC LAVA	VOLCANIC HIGHLAND MATERIALS	• CRUSTAL MATERIAL • VOLCANIC DEPOSITS
AGE	OLDER MARE FILLING	YOUNGER MARE FILLING	• EARLY HISTORY OF MOON • PRE-MARE MATERIAL • IMBRIUM BASIN FORMATION	• COMPOSITION AND AGE OF APENNINE FRONT MATERIAL • RILLE ORIGIN AND AGE • AGE OF IMBRIUM MARE FILL	• COMPOSITION AND AGE OF HIGHLAND CONSTRUCTION AND MODIFICATION • COMPOSITION AND AGE OF CAYLEY FORMATION	• COMPOSITION AND AGE OF HIGHLAND MASSIFS AND POSSIBLY OF LOW-LAND FILLING • COMPOSITION AND AGE OF DARK MANTLE • NATURE OF A ROCK LANDSLIDE

LAUNCH OPERATIONS

Prelaunch Preparations

NASA's John F. Kennedy Space Craft Center performs pre-flight checkout, test and launch of the Apollo 17 space vehicle. A government-industry team of about 600 will conduct the final countdown, 500 of them in Firing Room 1 in the Launch Control Center and 100 in the spacecraft control rooms in the Manned Spacecraft Operations Building (MSOB).

The firing room team is backed up by more than 5,000 persons who are directly involved in launch operations at KSC from the time the vehicle and spacecraft stages arrive at the Center until the launch is completed.

Initial checkout of the Apollo spacecraft is conducted in work stands and in the altitude chambers in the Manned Spacecraft Operations Building at Kennedy Space Center. After completion of checkout there, the assembled spacecraft is taken to the Vehicle Assembly Building (VAB) and mated with the launch vehicle. There the first integrated spacecraft and launch vehicle tests are conducted. The assembled space vehicle is then rolled out to the launch pad for final preparations and countdown to launch.

Flight hardware for Apollo 17 began arriving at KSC in October, 1970, while Apollo 14 was undergoing checkout in the VAB.

The command/service module arrived at KSC in late March, 1972, and was placed in an altitude chamber in the MSOB for systems tests and unmanned and manned chamber runs. During these runs, the chamber air was pumped out to simulate the vacuum of space at altitudes in excess of 200,000 feet. It is during these runs that spacecraft systems and astronauts' life support systems are tested.

The lunar module at KSC in June and its two stages were moved into an altitude chamber in the MSOB after an initial receiving inspection. It, too, was given a series of systems tests and unmanned and manned chamber runs. The prime and back-up crews participated in the chamber runs on both the LM and the CSM.

In July, the LM and CSM were removed from the chambers. After installing the landing gear on the LM and the SPS nozzle on the CSM, the LM was encapsulated in the spacecraft LM adapter (SLA) and the CSM was mated to the SLA. On August 24, the assembled spacecraft was moved to the VAB where it was mated to the launch vehicle.

The Lunar Roving Vehicle (LRV), which the Apollo 17 crew will use in their exploratory traverses of the lunar surface, arrived at KSC on June 2. Following a series of tests, which included a mission simulation on August 9 and a deployment demonstration on August 10, the LRV was flight installed in the Lunar Module's descent stage on August 13.

Erection of the Saturn V launch vehicle's three stages and instrument unit on Mobile Launcher 3 in the VAB's High Bay 3 began on May 15 and was completed on June 27. Tests were conducted on individual systems on each of the stages and on the overall launch vehicle before the spacecraft was erected atop the vehicle on August 24.

Rollout of the space vehicle from the VAB to Pad A at KSC's Launch Complex 39 was accomplished on August 28.

Processing and erection of the Skylab 1 and Skylab 2 launch vehicles was underway in the VAB while preparations were made for moving Apollo 17 to the pad, giving KSC three Saturn space vehicles "in flow" for the first time since the peak of Apollo activity in 1969.

After the move to the pad, the spacecraft and launch vehicle were electrically mated and the first overall test (plugs-in) was conducted on October 11.

The plugs-in test verified the compatibility of the space vehicle systems, ground support equipment, and off-site support facilities by demonstrating the ability of the systems to proceed through a simulated countdown, launch and flight. During the simulated flight portion of the test, the systems were required to respond to both normal and emergency flight conditions.

The space vehicle Flight Readiness Test was conducted October 18-20. Both the prime and backup crews participate in portions of the FRT, which is a final overall test of the space vehicle systems and ground support equipment when all systems are as near as possible to a launch configuration.

After hypergolic fuels were loaded aboard the space vehicle and the launch vehicle first stage fuel (RP-1) was brought aboard, the final major test of the space vehicle began. This was the Countdown Demonstration Test (CDDT), a dress rehearsal for the final countdown to launch.

The CDDT for Apollo 17 was divided into a "wet" and a "dry" portion. During the first or "wet" portion, the entire countdown, including propellant loading was carried out down to 8.9 seconds, the time for ignition sequence start. The astronaut crew did not participate in the wet CDDT.

At the completion of the wet CDDT, the cryogenic propellants (liquid oxygen and liquid hydrogen) were off-loaded and the final portion of the countdown was re-run, this time simulating the fueling and with the prime astronaut crew participating as they will on launch day.

Apollo 17 will mark the 12th Saturn V launch from KSC and the eleventh from Complex 39's Pad A. Only Apollo 10 was launched from Pad B.

Because of the complexity involved in the checkout of the 110.6 meter (363-foot) tall Apollo/Saturn V configuration, the launch teams make use of extensive automation in their checkout. Automation is one of the major differences in checkout used in Apollo compared to the procedures used in earlier Mercury and Gemini programs.

Computers, data display equipment and digital data techniques are used throughout the automatic checkout from the time the launch vehicle is erected in the VAB through liftoff. A similar but separate computer operation called ACE (Acceptance Checkout Equipment) is used to verify the flight readiness of the spacecraft. Spacecraft checkout is controlled from separate rooms in the MSOB.

COUNTDOWN

The Apollo 17 precount activities will start at T-6 days. The early tasks include electrical connections and pyrotechnic installation in the space vehicle. Mechanical buildup of the spacecraft is completed, followed by servicing of the various gases and cyrogenics to the CSM and LM. Once this is accomplished, the fuel cells are activated.

The final countdown begins at T-28 hours when the flight batteries are installed in the three stages and instrument unit of the launch vehicle.

At the T-9 hour mark, a built-in hold of nine hours and 53 minutes is planned to meet contingencies and provide a rest period for the launch crew. A one hour built-in hold is scheduled at T-3 hours 30 minutes.

Following are some of the highlights of the latter part of the count:

T-10 hours, 15 minutes	Start mobile service structure move to park site.
T-9 hours	Built-in hold for nine hours and 53 minutes. At end of hold, pad is cleared for LV propellant loading.
T-8 hours, 05 minutes	Launch vehicle propellant loading - Three stages (LOX in first stage, LOX and LH$_2$ in second and third stages). Continues thru T-3 hours 38 minutes.
T-4 hours, 00 minutes	Crew medical examination.
T-3 hours, 30 minutes	Crew supper.
T-3 hours, 30 minutes	One-hour built-in hold.
T-3 hours, 06 minutes	Crew departs Manned Spacecraft Operations Building for LC-39 via transfer van.
T-2 hours, 48 minutes	Crew arrival at LC-39
T-2 hours, 40 minutes	Start flight crew ingress

T-1 hour, 51 minutes	Space Vehicle Emergency Detection System test (Young participates along with launch team).
T-43 minutes	Retract Apollo access arm to standby position (12 degrees).
T-42 minutes	Arm launch escape system. Launch vehicle power transfer test, LM switch to internal power.
T-37 minutes	Final launch vehicle range safety checks (to 35 minutes)
T-30 minutes	Launch vehicle power transfer test, LM switch over to internal power.
T-20 minutes to T-10 minutes	Shutdown LM operational instrumentation.
T-15 minutes	Spacecraft to full internal power.
T-6 minutes	Space vehicle final status checks.
T-5 minutes, 30 seconds	Arm destruct system.
T-5 minutes	Apollo access arm fully retracted.
T-3 minutes, 6 seconds	Firing command (automatic sequence).
T-50 seconds	Launch vehicle transfer to internal power.
T-8.9 seconds	Ignition start.
T-2 seconds	All engines running.
T-0	Liftoff.

NOTE: Some changes in the countdown are possible as a result of experience gained in the countdown demonstration test which occurs about two weeks before launch.

-more-

Launch Windows

The mission planning considerations for the launch
phase of a lunar mission are, to a major extent, related to
launch windows. Launch windows are defined for two different
time periods: a "daily window" has a duration of a few hours
during a given 24-hour period: a "monthly window" consists
of a day or days which meet the mission operational constraints
during a given month or lunar cycle.

Launch windows are based on flight azimuth limits of
72° to 100° (Earth-fixed heading east of north of the launch
vehicle at the end of the roll program), on booster and spacecraft performance, on insertion tracking, and on Sun elevation
angle at the lunar landing site. All times are EST.

Launch Windows

LAUNCH DATE	OPEN	CLOSE	SUN ELEVATION ANGLE
December 6, 1972	9:53 pm EST	1:31 am	13°
December 7, 1972	9:53 pm	1:31 am	16.9-19.1°
January 4, 1973 *	9:51 pm EST	11:52 pm	6.8°
January 5, 1973	8:21 pm	11:51 pm	10.2-11.1°
January 6, 1973	8:28 pm	11:56 pm	20.3-22.4°
February 3, 1973	6:47 pm EST	10:13 pm	13.3-15.5°
February 4, 1973	6:58 pm	10:20 pm	13.5-15.5°

* Launch azimuth limits for January are 84° to 100°.

Ground Elapsed Time Update

It is planned to update, if necessary, the actual ground
elapsed time (GET) during the mission to allow the GET clock
to coincide with the preplanned major flight event times should
the event times be changed because of late liftoff or trajectory dispersions.

For example, if the flight plan calls for descent orbit
insertion (DOI) to occur at GET 88 hours, 55 minutes and the
flight time to the Moon is two minutes longer than planned
due to trajectory dispersions at translunar injection, the
GET clock will be turned back two minutes during the translunar coast period so that DOI occurs at the pre-planned
time rather than at 88 hours, 57 minutes. It follows that
the other major mission events would then also be accomplished
at the pre-planned GET times.

Updating the GET clock will accomplish in one adjustment what would otherwise require separate time adjustments for each event. By updating the GET clock, the astronauts and ground flight control personnel will be relieved of the burden of changing their checklists, flight plans, etc.

The planned times in the mission for updating GET will be kept to a minimum and will, generally, be limited to three updates. If required, they will occur at about 63, 96 and 210 hours into the mission. Both the actual GET and the update GET will be maintained in the MCC throughout the mission.

Synchronization of Ground Elapsed Time (GET)

The realtime GET is synchronized with the Flight Plan GET. In TLC, the GET is synchronized at 63:30 if the time propagated ahead to start of Rev 2 is more than +1 minute from the flight plan GET. In lunar orbit the GET is synchronized at 95:50 and 209:50 if the time propagated ahead to start of Rev 26 and Rev 66 respectively is more than +2 minutes from the flight plan GET. The synchronization is performed by a V70 uplink from the ground followed by the crew synchronizing the mission timer to the CMC clock.

LAUNCH AND MISSION PROFILE

The Saturn V launch vehicle (SA-512) will boost the Apollo 17 spacecraft from Launch Complex 39A at the Kennedy Space Center at 9:53 p.m. EST December 6, 1972, on an azimuth of 72 degrees.

The first stage (S-IC) will lift the vehicle 66 kilometers (33 nautical miles) above the Earth. After separation, the booster stage will fall into the Atlantic Ocean about 662 km (301 nm) downrange from Cape Kennedy approximately nine minutes, 12 seconds after liftoff.

The second stage (S-II) will push the vehicle to an altitude of about 173 km (87 nm). After separation, the S-II stage will follow a ballistic trajectory which will plunge it into the Atlantic about 4,185 km (2,093 nm) downrange about 19 minutes, 51 seconds into the mission.

The single engine of the third stage (S-IVB) will insert the spacecraft into a 173-kilometer (93 nm) circular Earth parking orbit before it is cut off for a coast period. When reignited, the engine will inject the Apollo spacecraft into a trans-lunar trajectory.

APOLLO 17
FLIGHT PROFILE

TRANSEARTH INJECTION (REV 75)
RENDEZVOUS
BRAKING (REV 52)
CSM 60 NM
CSM PLANE CHANGE (REV 48)
CSM ORBIT CIRCULARIZATION 62 NM (REV 12)
LM RCS DOI-2 7 X 60 NM (REV 13)
CSM/LM SEPARATION (REV 12)
CSM/LM DOI-1 15 X 59 NM (REV 3)
CSM/LM LOI ~51 X 171 NM

LM ASCENT (REV 51)
LM/CSM DOCKING (REV 52)
ASCENT STAGE JETTISON (REV 54)
LM LANDING
LM PDI (REV 13)
SIM DOOR JETTISON
S/C TRANSLUNAR TRAJECTORY
S-IVB IMPACT TRAJECTORY

IN-FLIGHT EVA
CSM TRANSEARTH TRAJECTORY
90-NM EARTH PARKING ORBIT
S-IVB APS EVASIVE MANEUVER

CM/SM SEPARATION
CM LANDING AND RECOVERY
LAUNCH
EARTH ORBIT INSERTION
S-IVB 2ND BURN CUTOFF ATLANTIC TRANSLUNAR INJECTION (TLI)
S/C SEPARATION, TRANSPOSITION, DOCKING & EJECTION

NOTE: NUMBERS ASSOCIATED WITH MAJOR LUNAR ORBIT EVENTS INDICATE ORDER OF OCCURRENCE.

8174

COMPARISON OF APOLLO MISSIONS

	PAYLOAD DELIVERED TO LUNAR SURFACE		EVA DURATION (HR:MIN)	SURFACE DISTANCE TRAVERSED (KM)	SAMPLES RETURNED	
	KG	(LBS)			KG	(LBS)
APOLLO 11	104	(225)	2:24	.25	20.7	(46)
APOLLO 12	166	(365)	7:29	2.0	34.1	(75)
APOLLO 14	209	(460)	9:23	3.3	42.8	(94)
APOLLO 15	550	(1210)	18:33	27.9	76.6	(169)
APOLLO 16	558	(1228)	20:14	26.7	95.4	(210)
APOLLO 17 (PLANNED)	558	(1228)	21:00	32.9	95.4	(210)

-more-

LAUNCH EVENTS

Time Hrs Min Sec	Event	Vehicle Wt Kilograms (Pounds)*	Altitude meters (Feet)*	Velocity Mtrs/Sec (Ft/Sec)*	Range Kilometers (Naut Mi)*
00 00 00	First Motion	2,923,461 (6,445,127)	60 (198)	0 (0)	0 (0)
00 01 23	Maximum Dynamic Pressure	1,823,015 (4,019,061)	13,329 (43,731)	505 (1,658)	6 (3)
00 02 19	S-IC Center Engine Cutoff	1,080,495 (2,382,085)	46,703 (153,224)	1,717 (5,634)	52 (28)
00 02 41	S-IC Outboard Engines Cutoff	842,718 (1,857,873)	66,498 (218,169)	2,365 (7,760)	92 (50)
00 02 43	S-IC/S-II Separation	675,991 (1,490,306)	68,197 (223,744)	2,371 (7,779)	96 (52)
00 02 45	S-II Ignition	675,991 (1,490,306)	69,727 (228,763)	2,365 (7,758)	99 (54)
00 03 13	S-II Aft Interstage Jettison	643,675 (1,419,060)	93,755 (307,597)	2,470 (8,103)	162 (88)
00 03 19	Launch Escape Tower Jettison	633,126 (1,395,804)	98,095 (321,835)	2,499 (8,198)	176 (95)
00 07 41	S-II Center Engine Cutoff	303,587 (669,295)	173,099 (567,910)	5,179 (16,992)	1,094 (591)
00 09 20	S-II Outboard Engines Cutoff	216,768 (477,893)	173,593 (569,532)	6,540 (21,458)	1,654 (893)
00 09 21	S-II/S-IVB Separation	171,158 (377,337)	173,634 (569,664)	6,534 (21,466)	1,661 (897)

*English measurements given in parentheses.

Time			Event	Vehicle Wt kilograms (Pounds)*	Altitude Meters (Feet)*	Velocity Mtrs/Sec (Ft/Sec)*	Range Kilometers (Naut Mi)*
00	09	24	S-IVB First Ignition	171,117 (377,337)	173,731 (569,984)	6,534 (21,467)	1,680 (907)
00	11	50	S-IVB First Cutoff	139,217 (306,921)	172,882 (567,199)	7,400 (24,278)	2,668 (1,440)
00	12	00	Parking Orbit Insertion	139,158 (306,791)	172,887 (567,215)	7,402 (24,284)	2,740 (1,479)
03	21	19	S-IVB Second Ignition	138,000 (304,237)	176,884 (580,326)	7,404 (24,291)	4,331 (2,339)
03	27	04	S-IVB Second Cutoff	65,156 (143,645)	298,545 (979,478)	10,440 (34,250)	7,228 (3,903)
03	27	14	Trans-Lunar Injection	65,092 (143,503)	312,029 (1,023,716)	10,432 (34,225)	7,326 (3,956)

*English measurements given in parentheses.

APOLLO 17
LUNAR ORBIT INSERTION

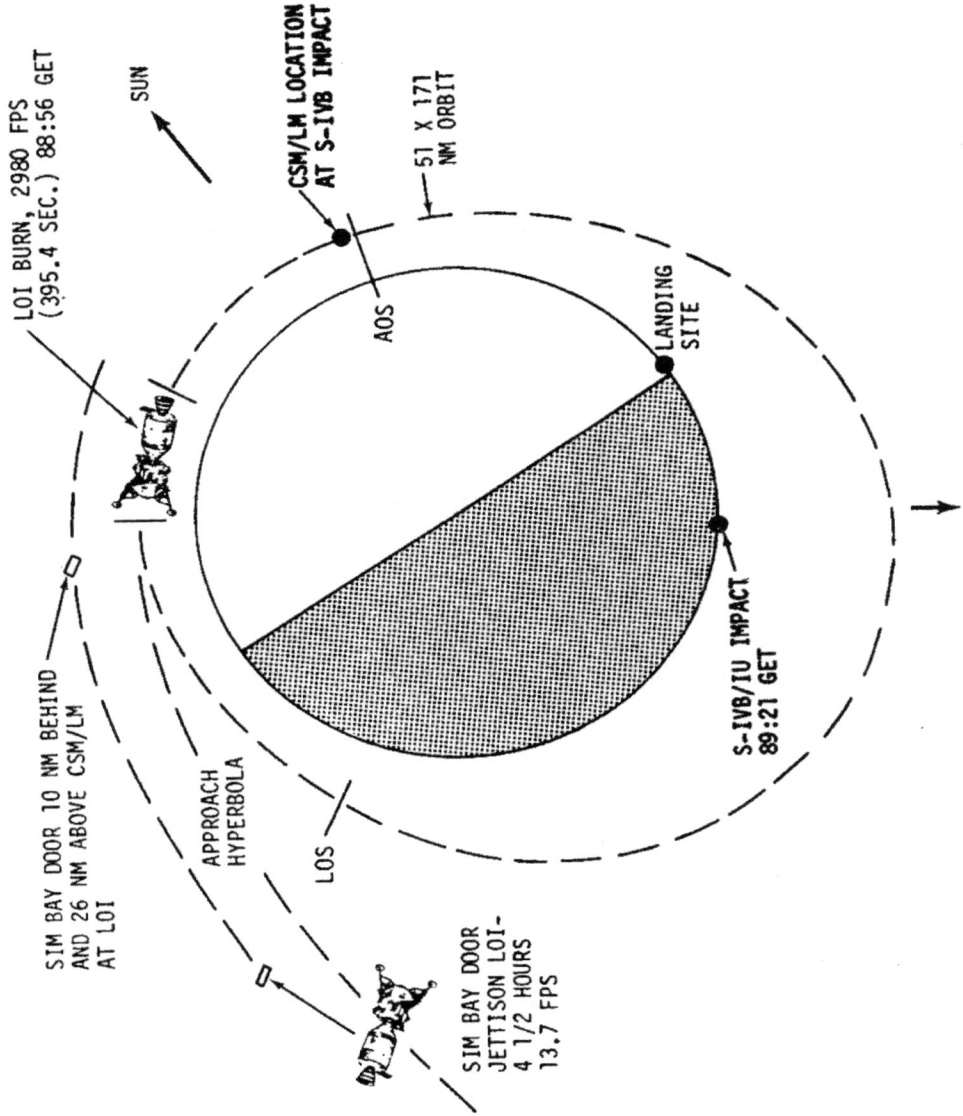

SUN

LOI BURN, 2980 FPS
(395.4 SEC.) 88:56 GET

CSM/LM LOCATION
AT S-IVB IMPACT

51 X 171
NM ORBIT

AOS

LANDING
SITE

SIM BAY DOOR 10 NM BEHIND
AND 26 NM ABOVE CSM/LM
AT LOI

APPROACH
HYPERBOLA

LOS

S-IVB/IU IMPACT
89:21 GET

SIM BAY DOOR
JETTISON LOI-
4 1/2 HOURS
13.7 FPS

APOLLO 17
CSM/LM LANDING EVENTS

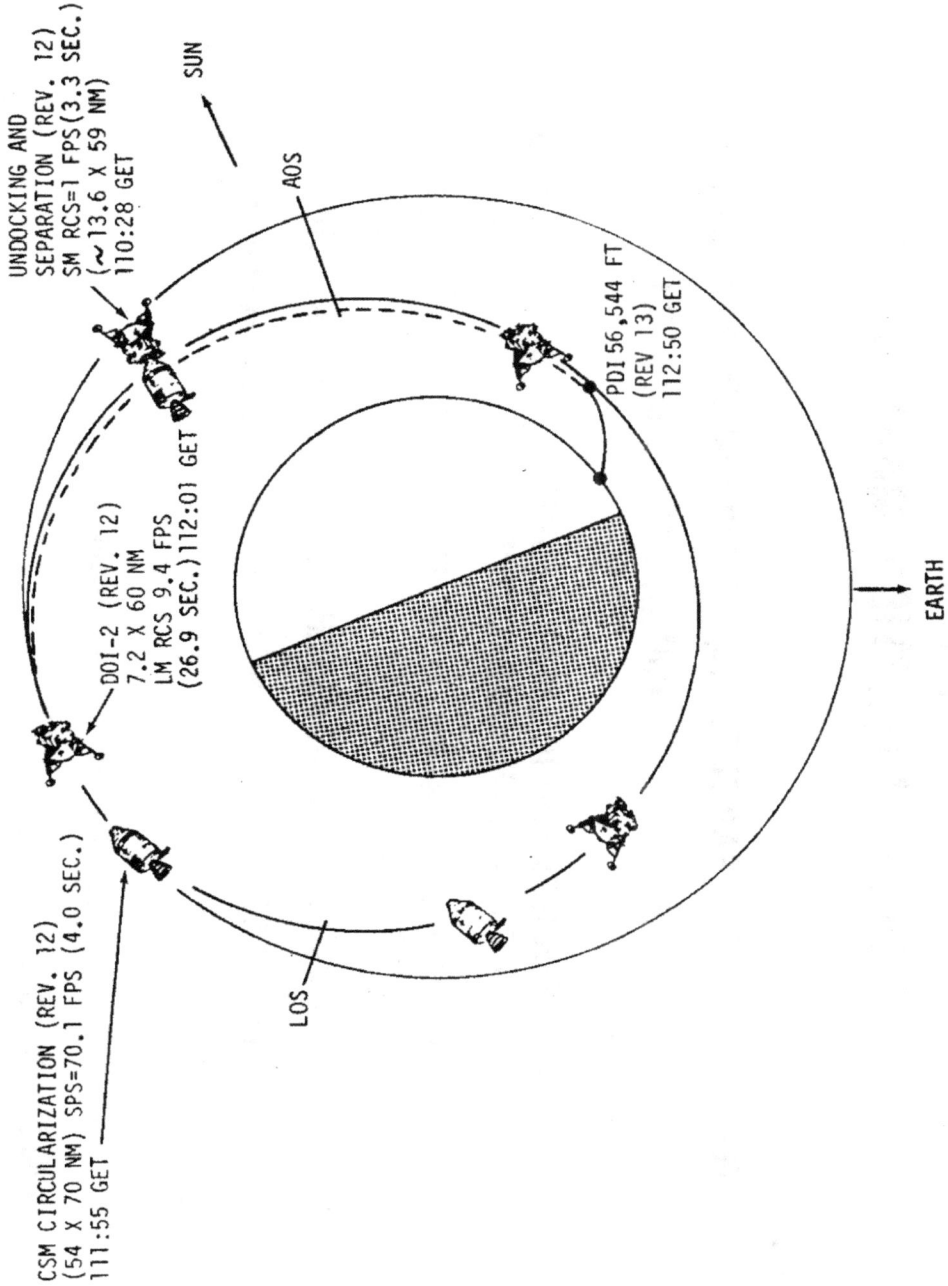

APOLLO 17
DESCENT ORBIT INSERTION MANEUVERS

● DOI-1

● PERILUNE ALTITUDE RAISED TO ~86,000 FT VS ~54,000 FT ON APOLLO 16

● PERILUNE LOCATION SHIFTED TO 10° W. OF LANDING SITE VS 16° E ON APOLLO 16

   (1) PROVIDES SUFFICIENT TIME FOR FLIGHT CONTROLLERS TO DETERMINE BURN CHARACTERISTICS

   (2) REDUCES PROBABILITY OF NECESSITY FOR DOI-1 BAILOUT MANEUVER

   (3) LANDMARK TRACKING ENHANCED BY HIGHER ALTITUDE

● SHOULD PRECLUDE EARLY CREW WAKEUP FOR A DOI TRIM MANEUVER

● DOI-2

● LOWERS PERILUNE FROM ~80,000 FT TO ~43,000 FT.

● 40 LBS OF LM RCS USED

● NET GAIN IN HOVER TIME OF ~3 SEC.

● SPS RESERVES INCREASED BY ~25 FPS

MISSION EVENTS

Events	GET hrs:min	Date/CST	Velocity change m/sec (ft/sec)	Purpose and resultant orbit
Translunar injection (S-IVB engine start)	3:27	7/12:20 am	3,048 (10,001)	Injection into translunar trajectory with 94km (51 nm) pericynthion
CSM separation, docking	4:02	7/12:55 am	---	Mating of CSM and LM
Ejection from SLA	4:47	7/1:40 am	.3 (1)	Separates CSM-LM from S-IVB/SLA
S-IVB evasive maneuver	5:10	7/2:03 am	3 (9.8)	Provides separation prior to S-IVB propellant and thruster maneuver to cause lunar impact
(S-IVB prop dump, APS burns from MSFC Launch Vehicle Trajectory documents)				
Midcourse correction 1	TLI+9 hr	7/9:20 am	0*	*These midcourse corrections have a nominal velocity change of 0 m/sec, but will be calculated in real time to correct TLI dispersions; trajectory remains within capability of a docked-DPS TEI burn should SPS fail to ignite.
Midcourse correction 2	TLI+32 hrs	8/8:20 am	0*	
Midcourse correction 3	LOI-22 hrs	9/3:48 pm	0*	
Midcourse correction 4	LOI-5 hrs	10/8:48 am	0*	
SIM door jettison	LOI-4.5 hrs	10/9:18 am	4.2 (13.7)	
Lunar orbit insertion	88:55	10/1:48 pm	-908.3 (-2980)	Inserts Apollo 17 into 94x316 km (51x170 nm) elliptical lunar orbit
S-IVB impacts lunar surface	89:21	10/2:14 pm		Seismic event for Apollo 12, 14, 15 and 16 passive seismometers. Target: 7 degrees south latitude by 8 degrees west longitude.
Descent orbit insertion No. 1	93:13	10/6:06 pm	-60.5 (-198.7)	SPS burn places CSM/LM into 25x109 km (15x59 nm) lunar orbit.
CSM/LM undocking	110:28	11/11:20 am	---	
CSM circularization burn	111:55	11/12:48 pm	21.4 (70.1)	Inserts CSM into 99.9x129.6 km (54x70 nm) orbit (SPS burn)

-27-

Events	GET hrs:min	Date/CST	Velocity change m/sec (ft/sec)	Purpose and resultant orbit
Descent orbit insertion No. 2	112:00	11/12:53 pm	2.8 (9.4)	Lowers LM pericynthion to 12.9 km (7 nm)
LM powered descent	112:49	11/1:42 pm	-1,850 (6,701)	Three-phase DPS burn to brake LM out of transfer orbit, vertical descent and lunar touchdown
LM lunar surface contact	113:01	11/1:54 pm	---	Lunar exploration, deploy ALSEP, collect geological samples, photography.
EVA-1 begins	116:40	11/5:33 pm	---	See separate EVA timelines
EVA-2 begins	139:10	12/4:03 pm	---	See separate EVA timelines
EVA-3 begins	162:40	13/3:33 pm	---	See separate EVA timelines
CSM plane change	182:36	14/11:29 am	102.6 (336.7)	Changes CSM orbital plane by 3.6 degrees to coincide with LM orbital plane at time of LM ascent.
LM ascent	188:03	14/4:56 pm	1,847.7 (6,062)	Boosts ascent stage into 16.6x88.5 km (9x47.8 nm) lunar orbit for rendezvous with CSM
Lunar orbit insertion	188:10	14/5:03 pm		
Terminal phase initiate (TPI) LM APS	188:57	14/5:50 pm	16.7 (54.8)	Boosts ascent stage into 88x118.5 km (47x64 nm) catch-up orbit; LM trails CSM by 59.2 km (32 nm) and 27.7 km (15 nm) below at TPI burn time.
Braking: 4 LM RCS burns	189:39	14/6:32 pm	9.5 (31.2)	Line-of-sight terminal phase braking to place LM in 114.8x114.8 km (62x62 nm) orbit for final approach, docking.
Docking	190:00	14/6:53 pm	---	CDR and LMP transfer back to CSM
LM jettison, separtion	194:09	14/11:02 pm	---	Prevents recontact of CSM with LM ascent stage for remainder of mission.
LM ascent stage deorbit	145:41	15/12:34 am	108.8 (-357)	ALSEP seismometers at Apollo landing sites record impact.

-more-

Events	GET hrs:min	Date/CST	Velocity change m/sec (ft/sec)	Purpose and resultant orbit
LM impact	195:58	15/12:51 am	---	Impact at about 1,643.9 m/sec (5,394 ft/sec) at -6.1 degree angle
Transearth injection (TEI)	236:39	16/5:32 pm	928.3 (3,045.7)	Injects CSM into transearth trajectory
Midcourse correction 5	253:40	17/1035 am	0	Transearth midcourse corrections will be computed in real time for entry corridor control and recovery area weather avoidance.
Transearth EVA	257:25	17/2:18 pm	---	Retrieve SM SIM bay film canisters.
Midcourse correction 6	EI-22 hrs	18/3:11 pm	0	
Midcourse correction 7	EI-3 hrs	19/10:11 am	0	
CM/SM separation (EI-15 min)	304:03	19/12:56 pm	---	Command module oriented for Earth atmosphere entry
Entry interface	304:18	19/1:11 pm		Command module enters Earth atmosphere at 11,000 m/sec (36,090 fps)
Splashdown	304:31	19/1:24 pm		Landing 2,111 km (1,140 nm) downrange from entry; splash at 17.9 degrees south latitude by 166 degrees west longitude.

POWERED DESCENT VEHICLE POSITIONS

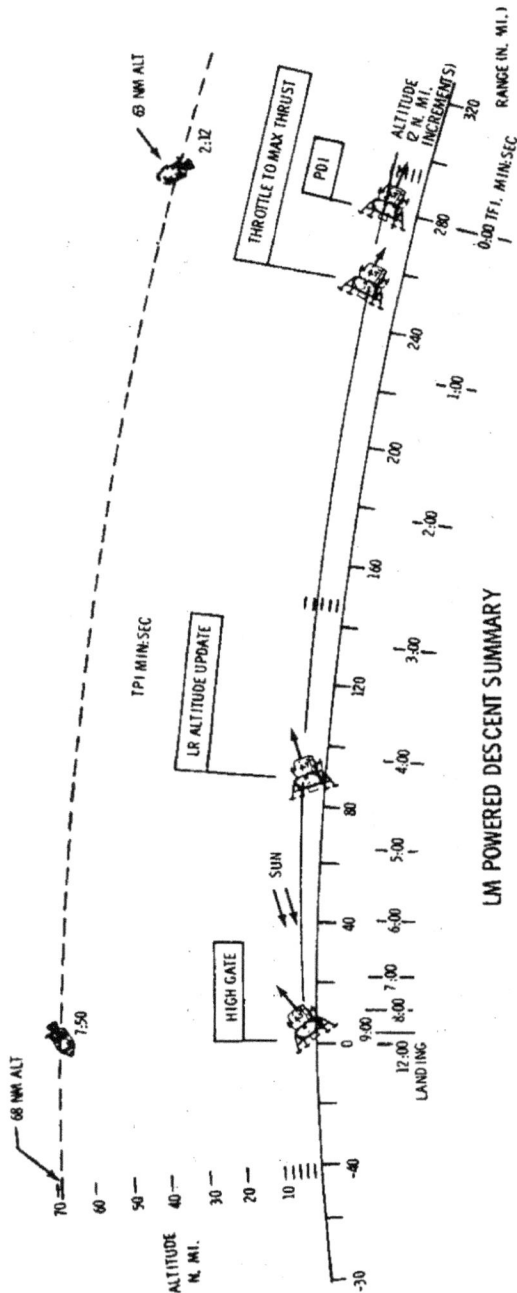

LM POWERED DESCENT SUMMARY

EVENT	TFI, MIN:SEC	V, FPS	Ḣ, FPS	H, FT.
POWERED DESCENT INITIATION	0:00	5568	-67	56,544
THROTTLE TO MAXIMUM THRUST	0:26	5542	-65	54,823
DPS THROTTLE RECOVERY	7:20	1202	-90	25,746
HIGH GATE	9:20	311	-177	8,159
LOW GATE	10:40	81	-25	709
LANDING	12:00	0	-5	6

APOLLO 17
APPROACH PHASE

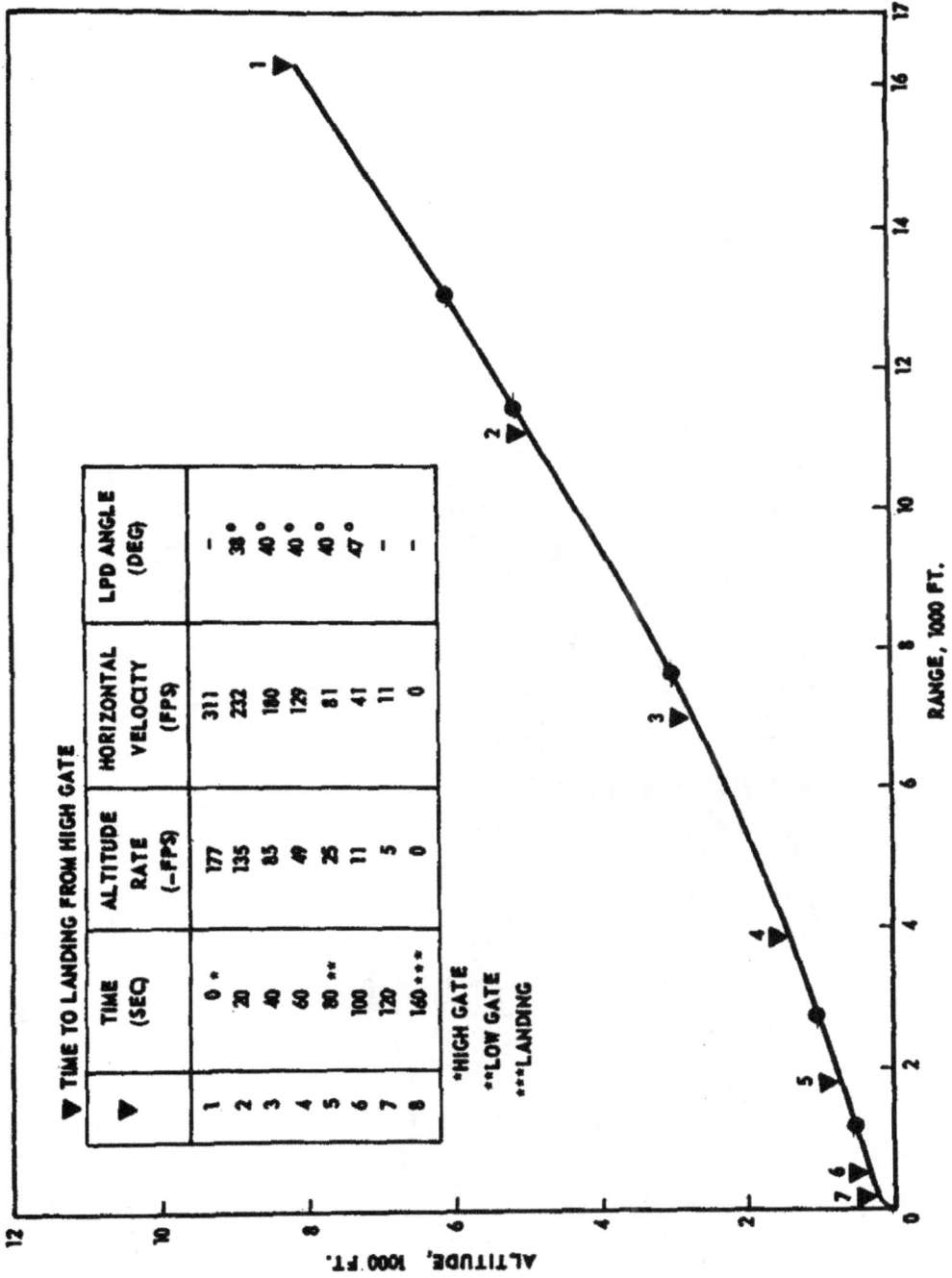

▼ TIME TO LANDING FROM HIGH GATE

▲	TIME (SEC)	ALTITUDE RATE (-FPS)	HORIZONTAL VELOCITY (FPS)	LPD ANGLE (DEG)
1	0 *	177	311	–
2	20	135	232	38°
3	40	85	180	40°
4	60	49	129	40°
5	80 **	25	81	40°
6	100	11	41	47°
7	120	5	11	–
8	160 ***	0	0	–

*HIGH GATE
**LOW GATE
***LANDING

ALTITUDE, 1000 FT.

RANGE, 1000 FT.

more›

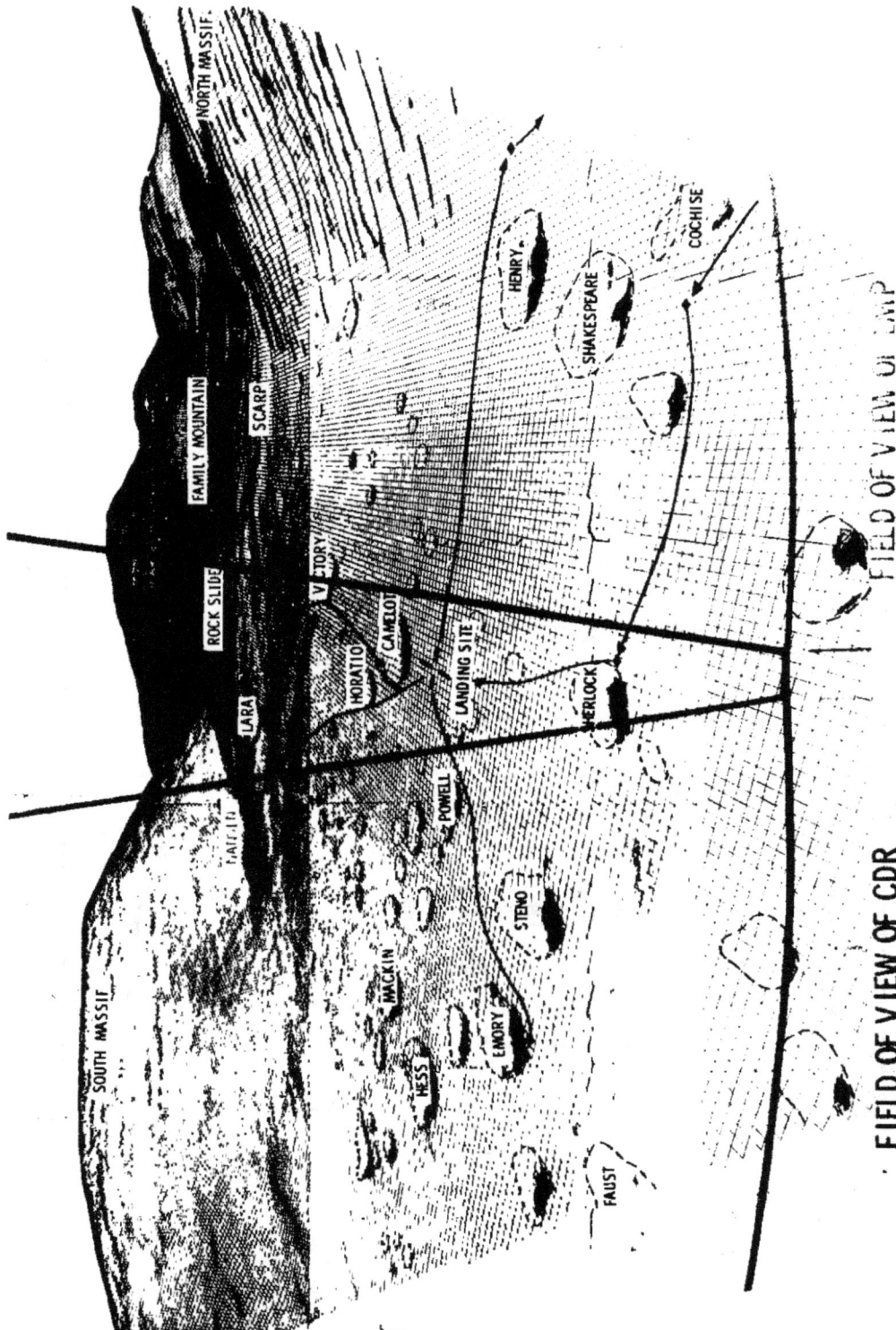

FIELD OF VIEW OF CDR

FIELD OF VIEW OF LMP

FILM RETRIEVAL FROM THE SIM BAY

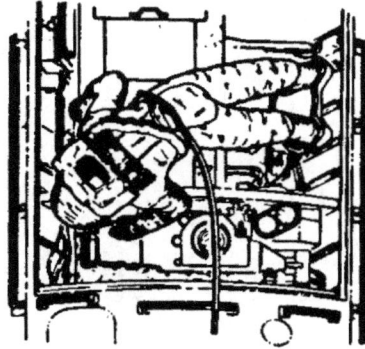

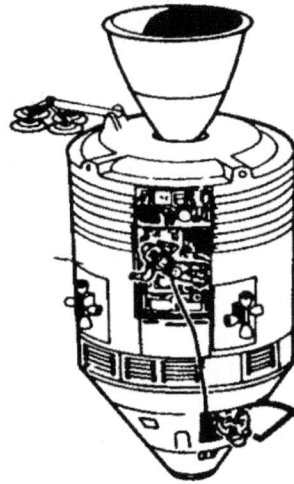

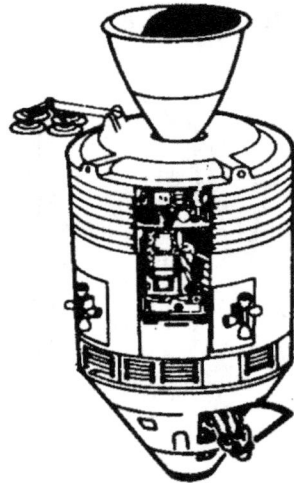

EVA TIMELINE

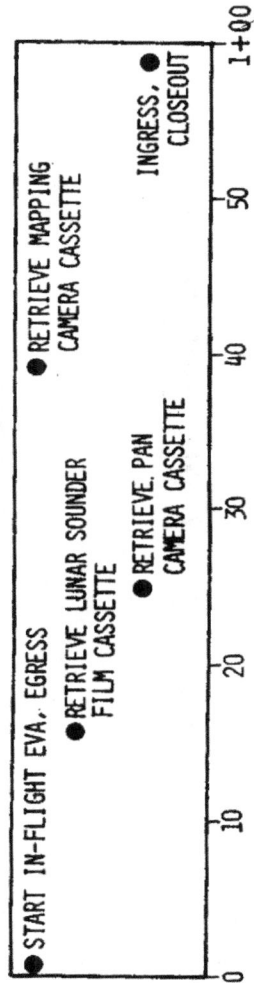

- START IN-FLIGHT EVA, EGRESS
- RETRIEVE LUNAR SOUNDER FILM CASSETTE
- RETRIEVE PAN CAMERA CASSETTE
- RETRIEVE MAPPING CAMERA CASSETTE
- INGRESS, CLOSEOUT

0   10   20   30   40   50   1+00

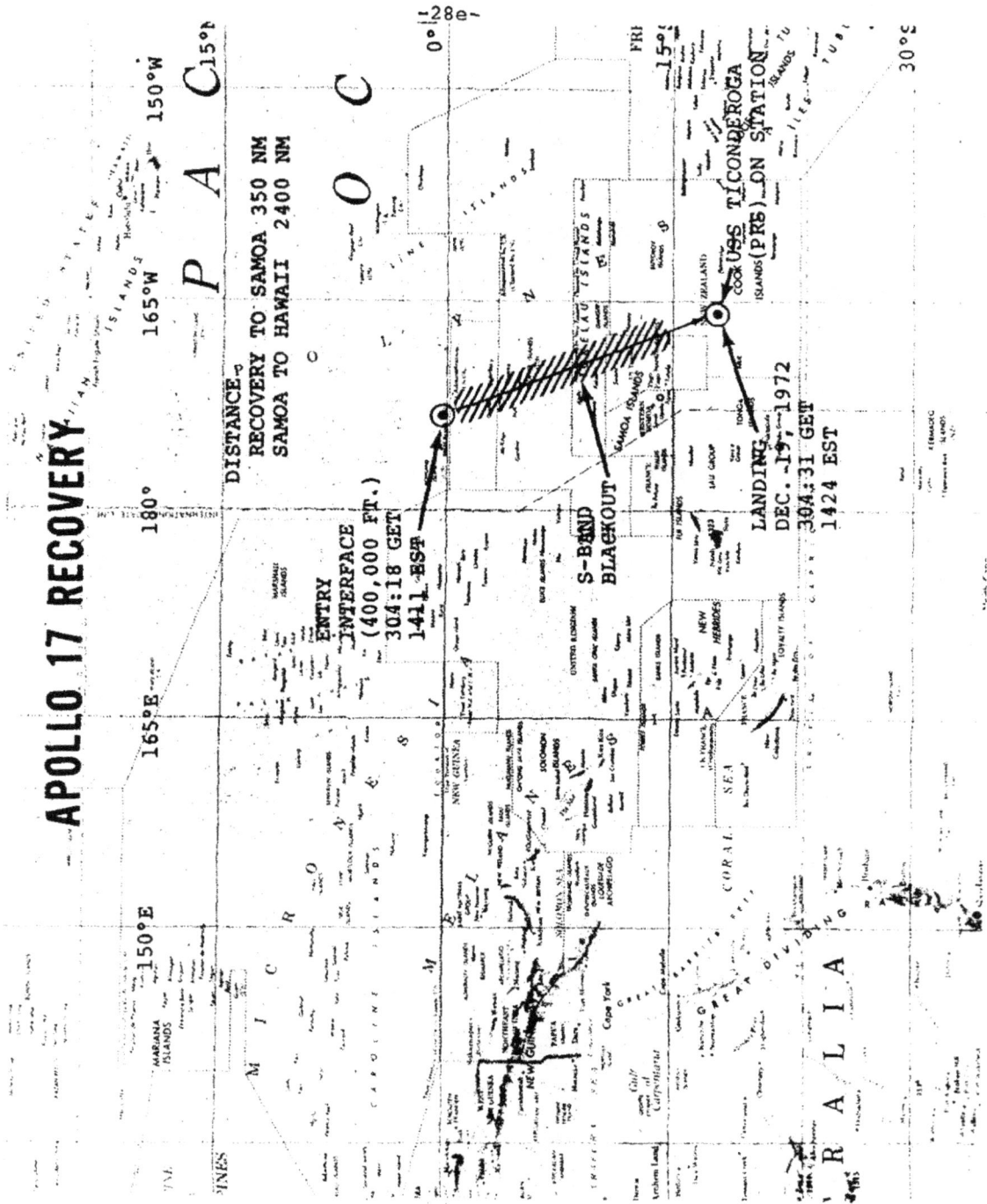

APOLLO 17 RECOVERY

DISTANCE
RECOVERY TO SAMOA 350 NM
SAMOA TO HAWAII 2400 NM

ENTRY
INTERFACE
(400,000 FT.)
304:18 GET
1411 EST

S-BAND
BLACKOUT

CO USS TICONDEROGA
(PR8) ON STATION

LANDING
DEC. 19, 1972
304:31 GET
1424 EST

APOLLO 17

CREW POST LANDING ACTIVITIES

DAYS FROM RECOVERY	DATE	ACTIVITY
SPLASHDOWN	DEC 19	
R + 1	DEC 20	DEPART SHIP, ARRIVE HAWAII
R + 2	DEC 21	DEPART HAWAII, ARRIVE HOUSTON
R + 3/4	DEC 22,23	CREW TECHNICAL DEBRIEFING PERIOD
	DEC 24 THRU JAN 2	NO DEBRIEFINGS SCHEDULED
	JAN 3	PICKUP CREW DEBRIEFINGS

EVA MISSION EVENTS

Events	GET hrs:min	Date/CST
Depressurize LM for EVA 1	116:40	11/5:33 pm
CDR steps onto surface	116:55	11/5:48 pm
LMP steps onto surface	116:58	11/5:51 pm
Crew offloads LRV	117:01	11/5:54 pm
CDR test drives LRV	117:20	11/6:13 pm
LRV parked near MESA	117:25	11/6:18 pm
LMP mounts geology pallet on LRV	117:29	11/6:22 pm
CDR mounts LCRV, TV on LRV	117:31	11/6:24 pm
LMP deploys United States flag	117:53	11/6:46 pm
CDR readies LRV for traverse	118:02	11/6:55 pm
LMP offloads ALSEP	118:07	11/8:00 pm
LMP carries ALSEP "barbell" to deployment site	118:20	11/7:13 pm
Crew begins ALSEP deploy	118:27	11/7:20 pm
ALSEP deploy complete	120:19	11/9:12 pm
Crew drive to Surface Electrical Properties (SEP) experiment site	120:38	11/9:31 pm
Crew arrives at SEP site and drops off transmitter	120:41	11/9:34 pm
Crew drives to station 1	120:46	11/9:39 pm
Enroute, crew emplaces Lunar Seismic Profiling Experiment (LSPE) explosive package	120:55	11/9:48 pm
Crew arrives station 1 for documented/rake/soil samples, crater sampling, trench etc.	121:07	11/10:00 pm
Crew emplaces LSPE explosive package No. 5 at station 1	122:05	11/10:58 pm

Event	GET hrs:min	Date/CST
Crew returns to SEP site	122:13	11/11:06 pm
LSPE explosive package No. 7 deployed enroute	122:27	11/11:20 pm
Crew arrives at SEP site	122:35	11/11:28 pm
SEP transmitter deploy completed	122:58	11/11:51 pm
Crew arrives at LM for EVA 1 closeout, sample packaging, load transfer bag	123:00	11/11:53 pm
LMP ingresses LM	123:24	12/12:17 am
CDR ingresses LM	123:36	12/12:29 am
Repressurize LM, end EVA 1	123:40	12/12:33 am
Depressurize LM for EVA 2	139:10	12/4:03 pm
CDR steps onto surface	139:26	12/4:19 pm
LMP steps onto surface	139:29	12/4:22 pm
Crew completes loading LRV for geology traverse	139:45	12/4:38 pm
Crewmen load geological gear on each other's PLSS	139:46	12/4:39 pm
LMP walks to SEP site, turns SEP on	139:50	12/4:43 pm
Crew drives toward station 2, deploys explosive package No. 4 enroute	140:02	12/4:55 pm
Crew arrives at station 2 for rake, core, documented samples and polarimetry at base of South Massif	141:08	12/5:01 pm
Crew drives toward station 3 collecting samples enroute with LRV sampling device	141:59	12/6:52 pm
Crew arrives at station 3 for rake, trench, and documented sampling of scarp and light mantle	142:28	12/7:21 pm

Event	GET hrs:min	Date/CST
Crew drives toward station 4	143:13	12/8:06 pm
Crew arrives at station 4 for observations, rake and documented samples and a double core around dark halo crater	143:32	12:8:25 pm
Crew drives toward station 5, deploys explosive package No. 1 enroute	144:13	12/9:06 pm
Crew arrives at station 5 for observations, double core, rake and documented samples around 700-meter mantled crater	144:46	12/9:39 pm
Crew drives back to LM, deploys explosive package No. 8 enroute	145:16	12/10:09 pm
Crew arrives at LM for EVA closeout, packages samples, film mags	145:26	12/10:19 pm
LMP ingresses LM	145:57	12/10:50 pm
CDR ingresses LM	146:06	12/10:59 pm
Repressurize LM, end EVA 2	146:10	12/11:03 pm
Depressurize LM for EVA 3	162:40	13/3:33 pm
CDR steps onto surface	162:55	13/3:48 pm
LMP steps onto surface	162:59	13/3:52 pm
Crew completes loading LRV for geology traverse	163:06	13/3:59 pm
Crewmen load geological gear on each other's PLSS	163:10	13/4:03 pm
LMP walks to SEP site, turns SEP on	163:15	13/4:08 pm
CDR drives to SEP site, picks up LMP, depart for station 6	163:25	13/4:18 pm

-more-

Events	GET hrs:min	Date/CST
Crew arrives at station 6 for rake and documented samples and polarimetry near base of North Massif	163:52	13/4:45 pm
Crew drives to station 7	164:39	13/5:32 pm
Crew arrives at station 7 for rake and documented sampling at base of North Massif	164:50	13/5:43 pm
Crew drives to station 8	165:37	13/6:30 pm
Crew arrives at station 8 for rake and documented samples at base of sculptured hills	165:50	13/6:43 pm
Crew drives to station 9	166:37	13/7:30 pm
Crew arrives at station 9 for radial, rake and documented samples at fresh 80-meter crater on dark mantle	166:53	13/7:46 pm
Crew drives to station 10	167:23	13/8:16 pm
Crew arrives at station 10, for documented samples, double core, on rim of blocky-rimmed crater	167:47	13/8:40 pm
Crew drives back to LM, deploys explosive package No. 2 enroute	168:23	13/9:16 pm
Crew arrives at LM and crew begins EVA 3 close-out, sample stowage, etc.	168:42	13/9:35 pm
LMP hikes to ALSEP site to fetch neutron probe	168:59	13/9:52 pm
CDR deploys explosive package No. 3 near LRV final parking location	169:21	13/10:14 pm
LMP ingresses LM	169:32	13/10:25 pm
CDR ingresses LM	169:38	13/10:31 pm
Repressurize LM, end EVA 3	169:40	13/10:33 pm

-more-

APOLLO 17 VS APOLLO 16
OPERATIONAL DIFFERENCES

ITEM	APOLLO 17	APOLLO 16
LAUNCH TIME	NIGHT	DAY
TRANSLUNAR INJECTION	ATLANTIC (3RD REV)	PACIFIC (2ND REV)
DESCENT ORBIT INSERTION	DOI-1 & 2 MANEUVERS	ONE DOI MANEUVER
PERILUNE LOCATION	10° W OF LANDING SITE	16° E OF LANDING SITE
LUNAR SURFACE STAY	75 HOURS	73 HOURS (PLANNED) 71 HOURS (ACTUAL)
TRAVERSE DISTANCE	32.9 KM	25.2KM (PLANNED) 26.7 KM (ACTUAL)
LUNAR ORBIT PLANE CHANGES	1	2 (PLANNED)
EARTH RETURN INCLINATION	66.5° DESCENDING	62° ASCENDING
TOTAL MISSION TIME	304:31 (PLANNED)	290:36 (PLANNED) 265:51 (ACTUAL)

APOLLO 17 LRV TRAVERSES

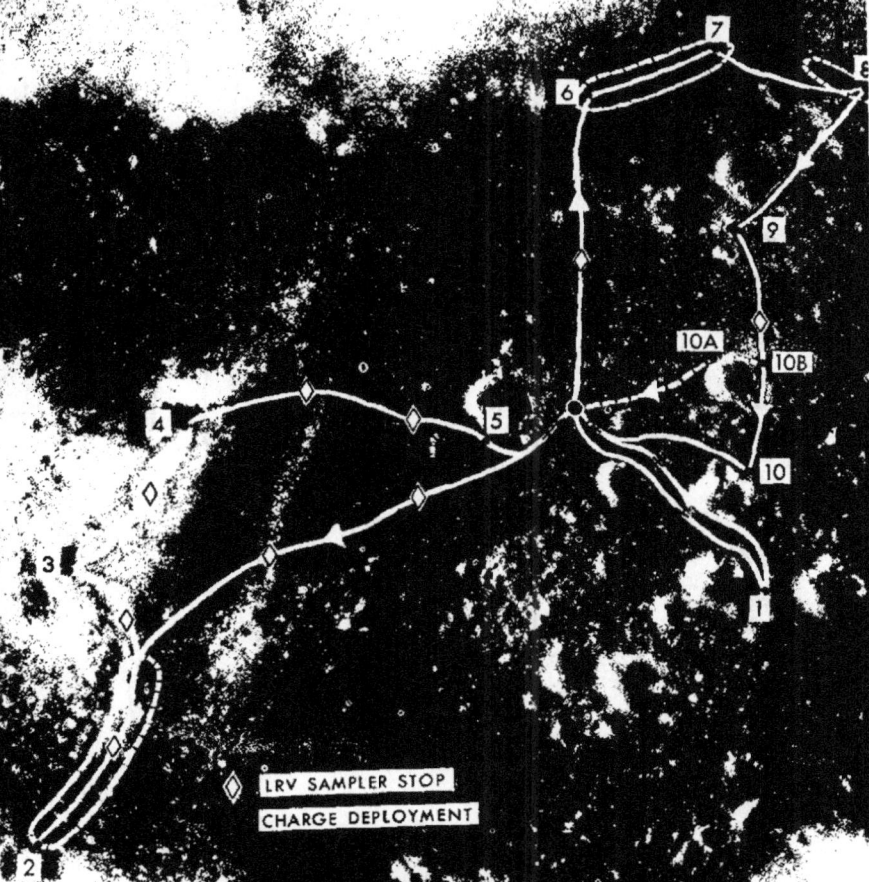

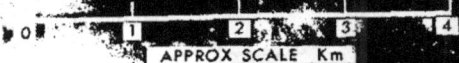

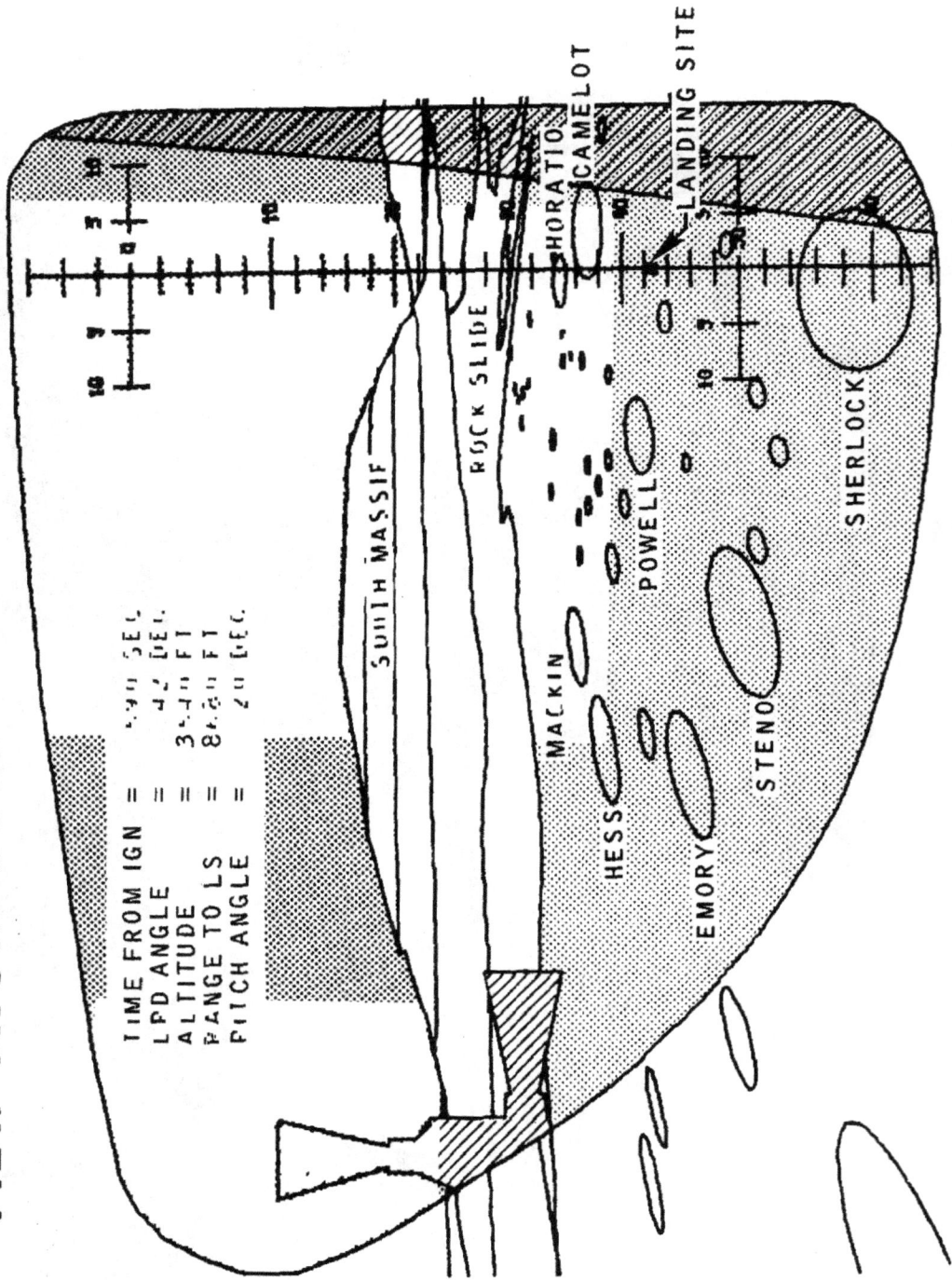

VIEW FROM COMMANDER'S WINDOW

MPAD 72-428-S

TIME FROM IGN = ... SEC.
LPD ANGLE = ... DEG.
ALTITUDE = 3... FT
RANGE TO LS = 8... FT
PITCH ANGLE = ... DEG.

SOUTH MASSIF

ROCK SLIDE

HORATIO

CAMELOT

LANDING SITE

MACKIN

HESS

EMORY

POWELL

STENO

SHERLOCK

APOLLO 17 LUNAR SURFACE TIMELINE

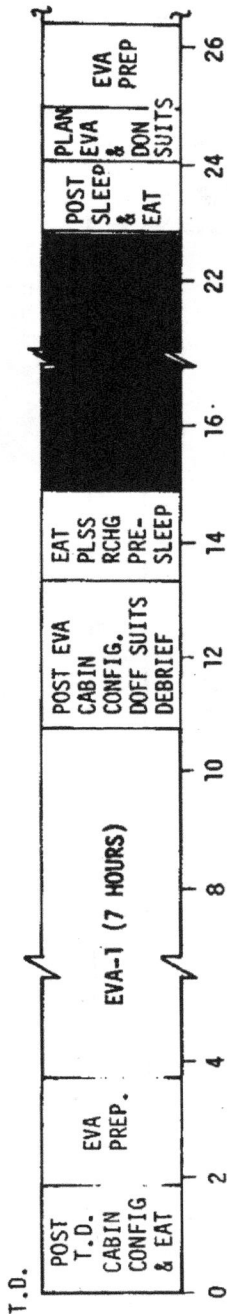

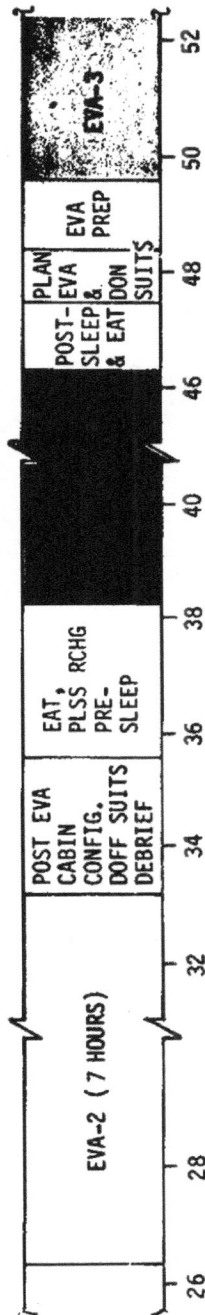

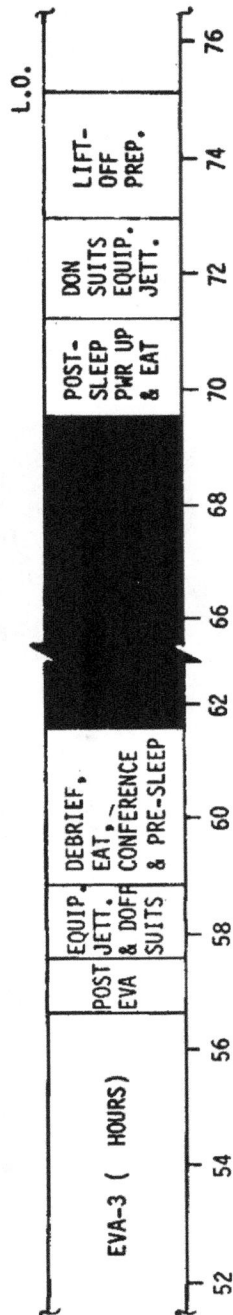

APOLLO 17

TRAVERSE STATION TIMELINE - EVA 1

STATION 1: EMORY (1:06)

CDR	OVERHEAD	DESCRIPTION	SAMPLING	O/H
	:05	:05	:52	:04

LMP	O/H	PAN	DESCRIPTION	RAKE/SOIL SAMPLE AND SAMPLING	O/H

NOTES:

O/H = OVERHEAD

APOLLO 17

TRAVERSE STATION TIMELINE - EVA 2

STATION 2: NANSEN (:51)

CDR	O/H	DESCRIPTION AND SAMPLING	O/H
	:05	:21	:04
LMP	O/H	RAKE/SOIL SAMPLE/POLARIMETRY	RAKE SAMPLE, SAMPLING AND SINGLE CORE
	:05		:16
PAN			

STATION 3: LARA (:45)

CDR	O/H	DESCRIPTION AND SAMPLING	SAMPLING
	:05	:13	
LMP	O/H	RAKE/SOIL SAMPLING	SAMPLING AND EXPLORATORY TRENCH
	:05		:18
PAN			O/H
			:04
			O/H

STATION 4: SHORTY (:41)

CDR	O/H	DESCRIPTION AND SAMPLING	O/H
	:05	0:11	:04
LMP	O/H	RAKE/SOIL SAMPLING	DOUBLE CORE	O/H
	:05		0:16
PAN			

STATION 5: CAMELOT (:30)

CDR	O/H	DESCRIPTION AND SAMPLING	O/H
	:05	:11	:04
LMP	O/H	PAN	RAKE/SOIL SAMPLING	DOUBLE CORE	O/H
	:05			

-more-

APOLLO 17

TRAVERSE STATION TIMELINE - EVA 3

STATION 6 & 7: N. MASSIF AREA (1:28)

CDR	O/H :10	DESCRIPTION :10	SAMPLING :20	SAMPLING :40	PAN AND O/H :08
LMP	O/H	PAN	DESCRIPTION :10	RAKE/SOIL SAMPLING POLARIMETRY	SAMPLING	PAN AND O/H :08

STATION 8: SCULPTURED HILLS AREA (:44)

CDR	O/H :05	DESCRIPTION :05	SAMPLING :20	O/H :04
LMP	O/H	PAN	DESCRIPTION	RAKE/SOIL SAMPLE	SAMPLING	O/H

STATION 9: VAN SERG (:29)

CDR	O/H :05	DESCRIPTION :05	SAMPLING :15	O/H :04
LMP	O/H	PAN	DESCRIPTION	RAKE/SOIL SAMPLE	O/H

STATION 10: SHERLOCK (:36)

CDR	O/H :05	DESCRIPTION :05	SAMPLING :12	O/H :04
LMP	O/H	PAN	DESCRIPTION	SAMPLING	DOUBLE CORE :10	O/H

APOLLO 17
EVA 1 TIMELINE

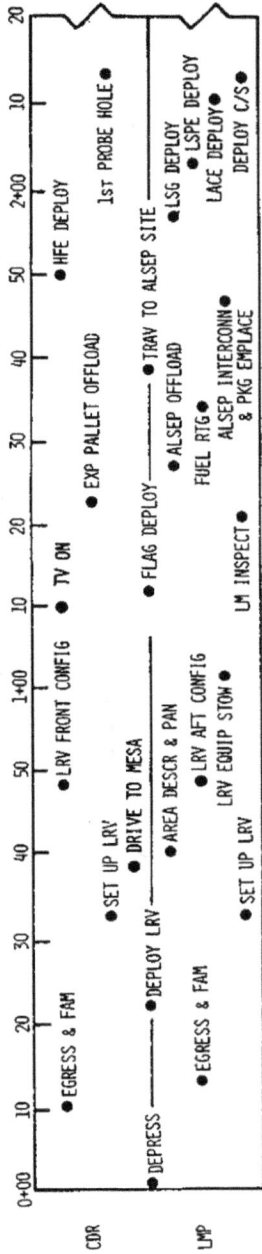

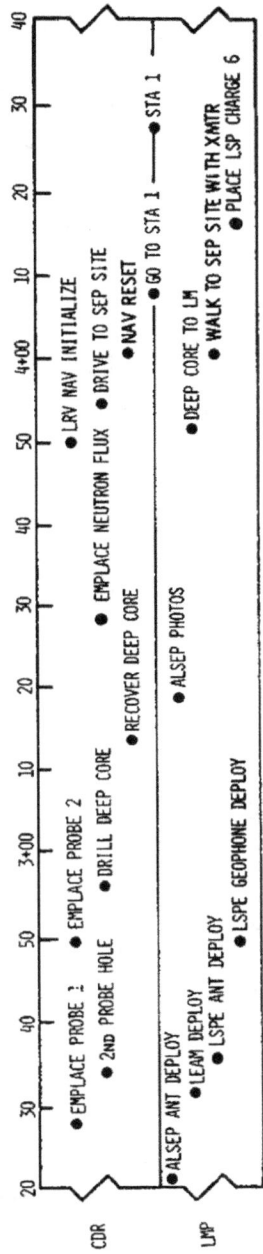

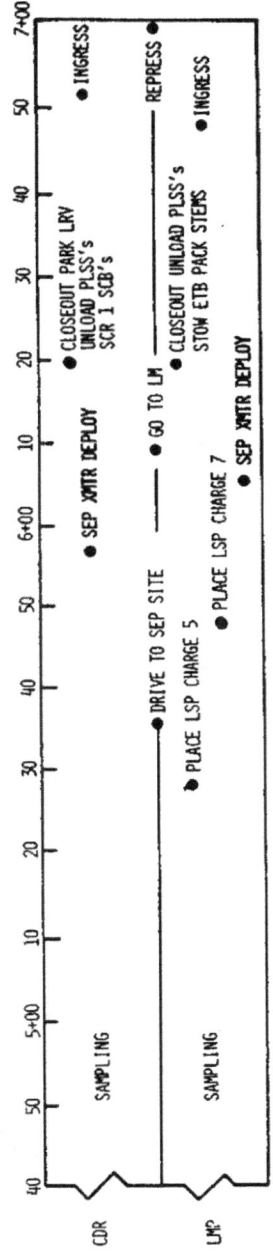

APOLLO 17
EVA 2 TIMELINE

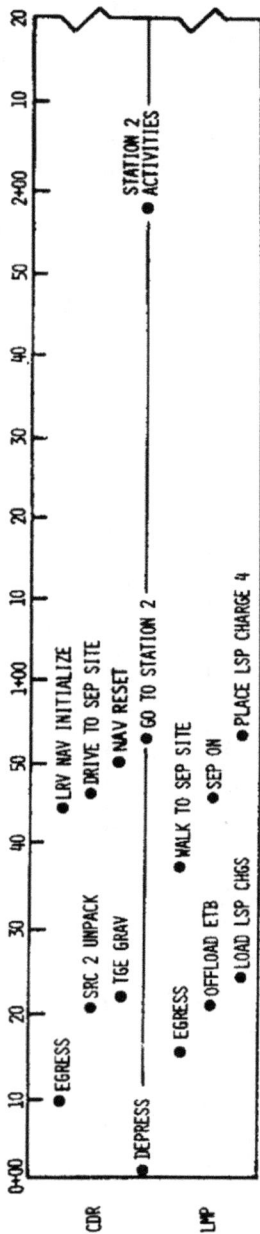

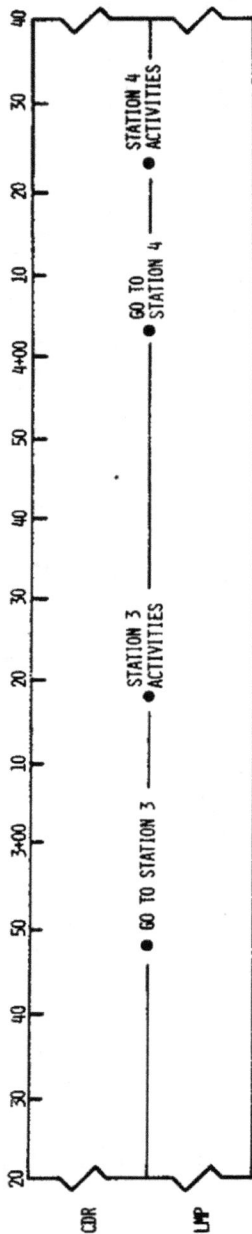

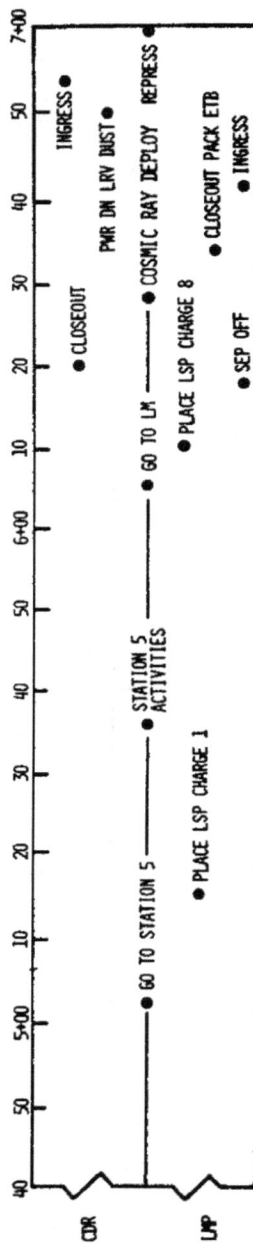

APOLLO 17
EVA 3 TIMELINE

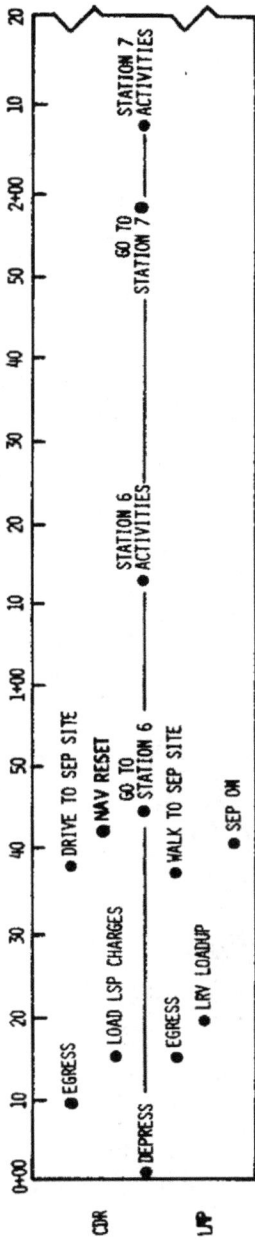

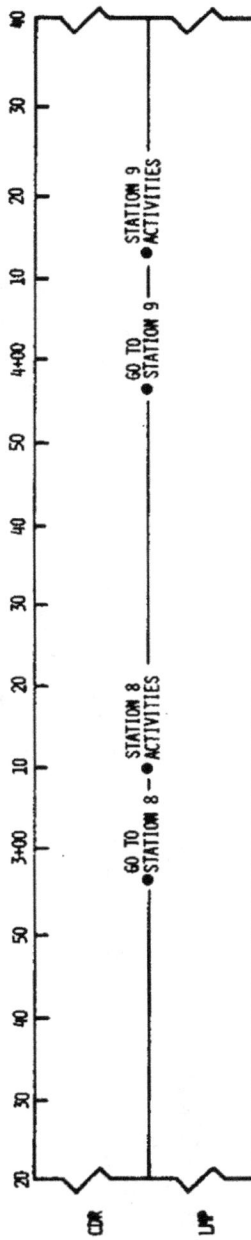

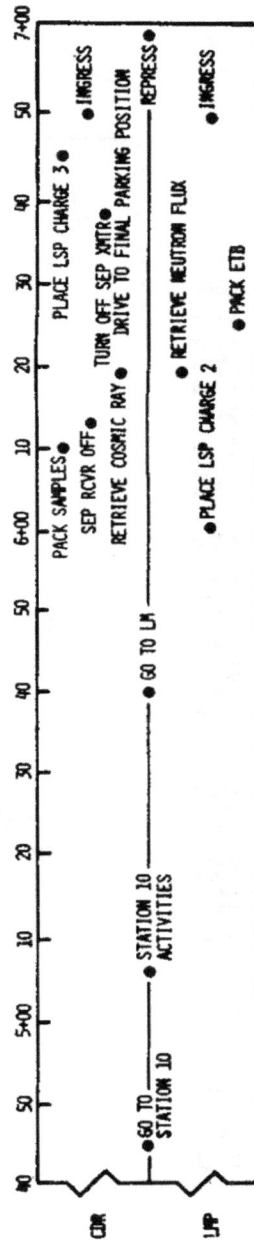

APOLLO 17

NEAR LM LUNAR SURFACE ACTIVITY

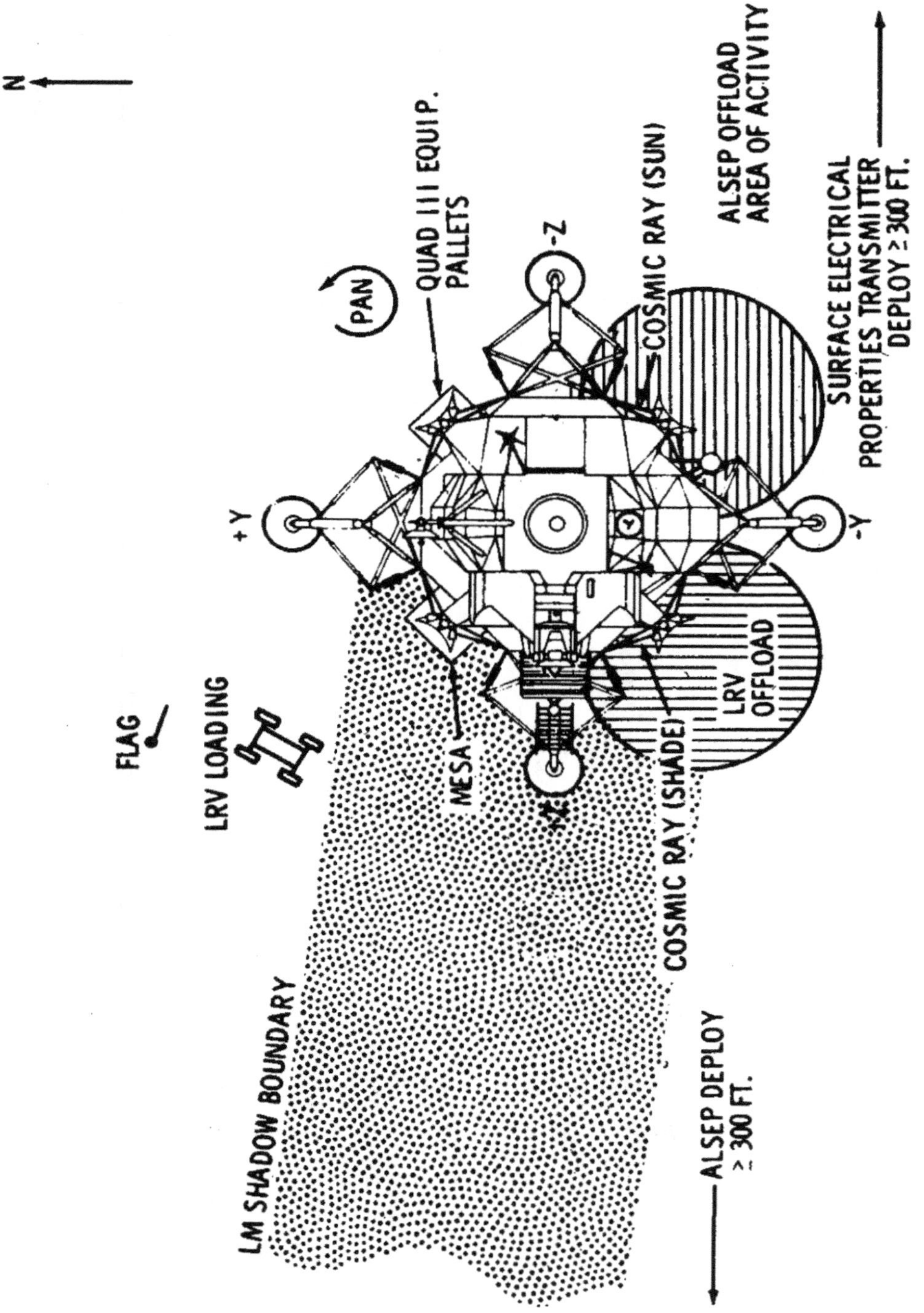

N

FLAG

LRV LOADING

LM SHADOW BOUNDARY

ALSEP DEPLOY ≥ 300 FT.

COSMIC RAY (SHADE)

MESA

PAN

QUAD III EQUIP. PALLETS

+Y

-Y

-Z

+Z

COSMIC RAY (SUN)

LRV OFFLOAD

SURFACE ELECTRICAL PROPERTIES TRANSMITTER DEPLOY ≥ 300 FT.

ALSEP OFFLOAD AREA OF ACTIVITY

TAURUS-LITTROW -- APOLLO 17 LANDING SITE

Landing site for the final Apollo lunar landing mission, Taurus-Littrow, takes its name from the Taurus mountains and Littrow crater which are located in a mountainous region on the southeastern rim of the Serenitatis basin.

The actual target landing site is at 30° 44'58.3" east longitude by 20°09'50.5" north latitude --- about 750 km east of the Apollo 15 landing site at Hadley Rille.

Geologists speculate that most of the landing site region probably consists of highland material which was uplifted to its present height at the time the Serenitatis basin was formed. The valley in which the landing site is located is covered by a fine-grained dark mantle that may consist of volcanic fragments. The site is surrounded by three high, steep massifs which likely are composed of breccia formed by the impacts that created some of the major mare basins --- probably pre-Imbrian in age.

A range of "sculptured hills" to the northeast of the landing site is believed to be of the same origin as the massifs, but probably having a different history of erosion and deformation. In gross morphology, the sculptured hills possess some of the characteristics of volcanic structures.

Most of the plain between the massifs is covered by a dark mantle which apparently has no large blocks or boulders, and which has been interpreted to be a pyroclastic deposit. The dark mantle is pocked by several small, dark halo craters that could be volcanic vents all near the landing site. The dark mantle material is thought to be younger in age than all of the large craters on the plain --- probably Eratusthenian Copernican age.

Extending northward from the south massif is a bright mantle with ray-like fingers which overlies the dark mantle. Geologists believe the light mantle is from an avalanche of debris down the slopes of the south massif, and that, it is of Copernican age.

Craters near the landing site range from large, steep-side craters-one-half to one kilometer in diameter-that are grouped near the landing point to scattered clusters of craters less than a half-kilometer across.

Another prominent landing site feature is an 80-meter high scarp trending roughly north-south near the west side of the valley into the north massif. The scarp is thought to be a surface expression of a fault running through the general region.

-more-

The Apollo 17 lunar module will approach from the east over the 750 meter hills, clearing then by about 3000 meters.

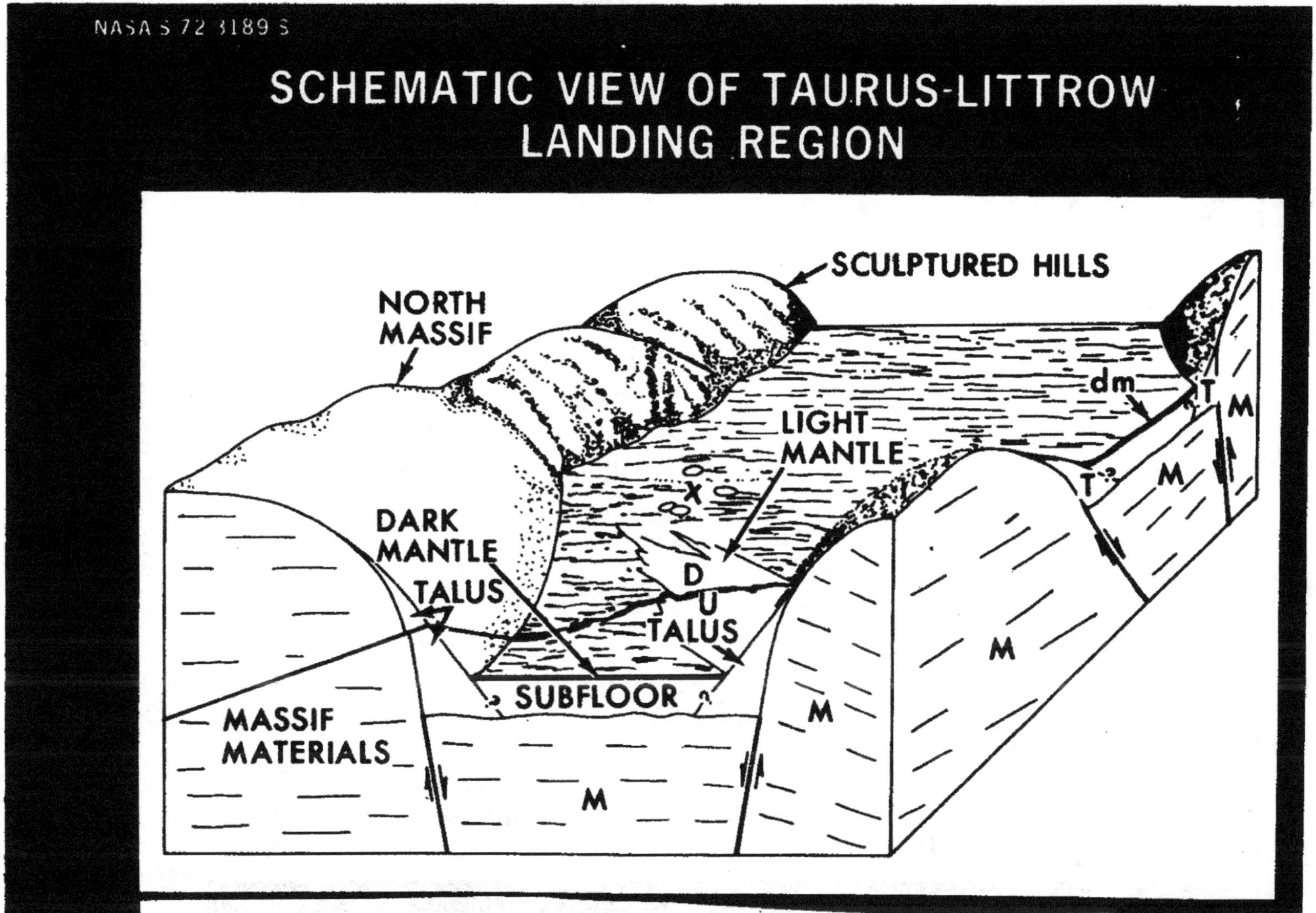

SCHEMATIC VIEW OF TAURUS-LITTROW LANDING REGION

Fig. 8 - Schematic view of Taurus-Littrow landing and traverse region looking east-southeast. X is nominal landing site. Vertical exaggeration about 2.5.

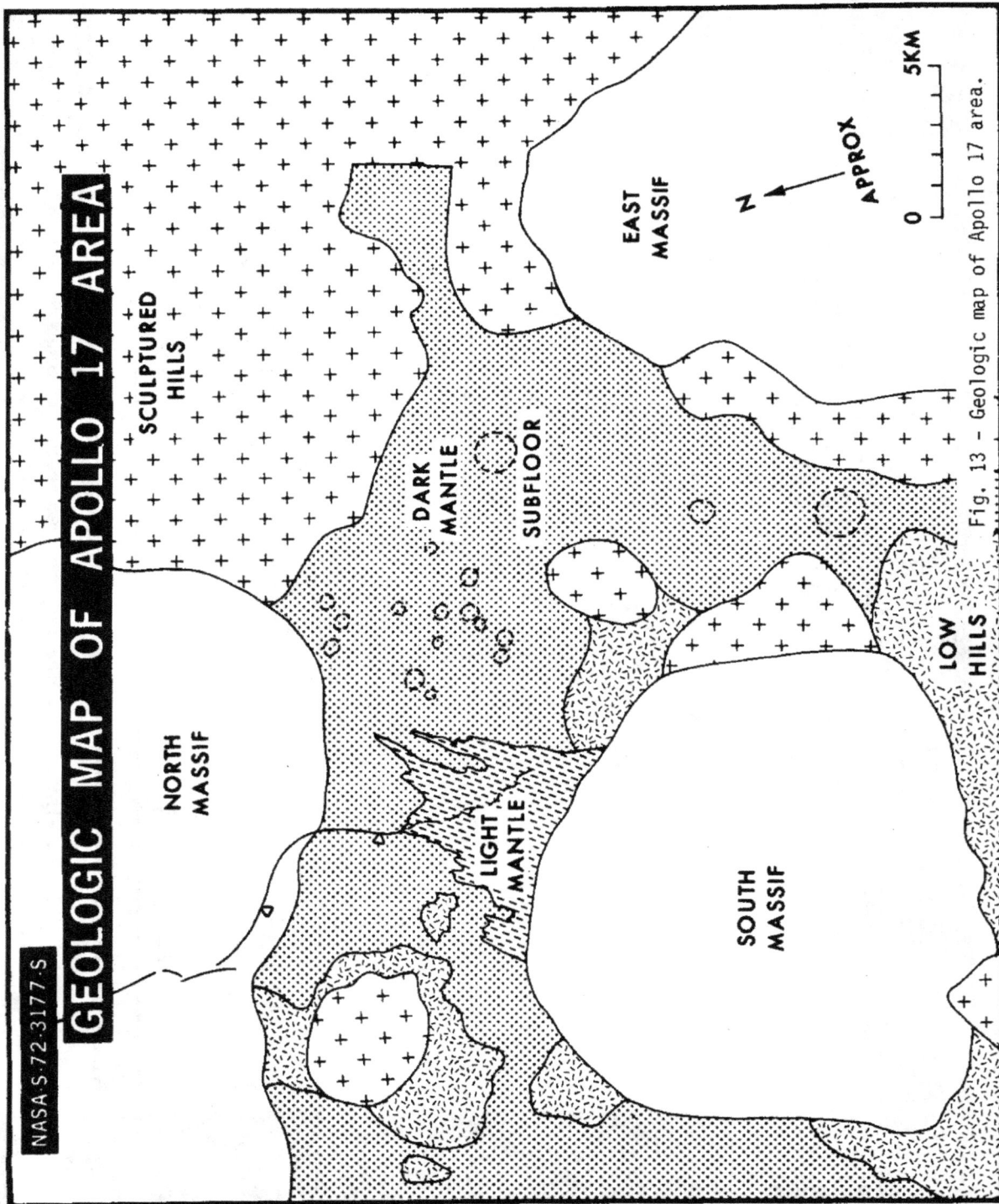

GEOLOGIC MAP OF APOLLO 17 AREA

NASA·S·72·3177·S

NORTH MASSIF

SCULPTURED HILLS

DARK MANTLE

SUBFLOOR

EAST MASSIF

LIGHT MANTLE

SOUTH MASSIF

LOW HILLS

N

APPROX

0   5KM

Fig. 13 - Geologic map of Apollo 17 area.

LUNAR SURFACE SCIENCE

S-IVB Lunar Impact

The Saturn V's third stage, after it has completed its job of placing the Apollo 17 spacecraft on a lunar trajectory will be aimed to impact on the Moon. As in several previous missions, this will stimulate the passive seismometers left on the lunar surface in earlier Apollo flights.

The S-IVB, with instrument unit attached, will be commanded to hit the Moon 305 kilometers (165 nautical miles) southeast of the Apollo 14 ALSEP site, at a target point 7 degrees south latitude by 8 degrees west longitude, near Ptolemaeus Crater.

After the spacecraft is ejected from the launch vehicle, a launch vehicle auxiliary propulsion system (APS) ullage motor will be fired to separate the vehicle a safe distance from the spacecraft. Residual liquid oxygen in the almost spent S-IVB/IU will then be dumped through the engine with the vehicle positioned so the dump will slow it into an impact trajectory. Mid-course corrections will be made with the stage's APS ullage motors if necessary.

The S-IVB/IU will weigh 13,931 kilograms (30,712 pounds) and will be traveling 9,147 kilometers an hour (4,939 nautical mph) at lunar impact. It will provide an energy source at impact equivalent to about 11 tons of TNT.

ALSEP Package

The Apollo Lunar Surface Experiments Package (ALSEP) array carried on Apollo 17 has five experiments: heat flow, lunar ejecta and meteorites, lunar seismic profiling, lunar atmospheric composition and lunar surface gravimeter.

Additional experiments and investigations to be conducted at the Taurus-Littrow landing site will include traverse gravimeter, surface electrical properties, lunar neutron probe, cosmic ray director, soil mechanics and lunar geology investigation.

S-IVB/IU IMPACT

SEA OF SERENITY

SEA OF TRANQUILITY

SEA OF NECTAR

TAURUS LITTROW APOLLO 17

APOLLO 11

APOLLO 16

DESCARTES

SEA OF VAPORS

HADLEY RILLE APOLLO 15

CENTRAL BAY

ALPHONSUS

APPENNINE MOUNTAIN

SEETHING BAY

RAINS

APOLLO 15 S-IVB IMPACT

APOLLO 14 FRA MAURO

TARGET APOLLO 17 S-IVB IMPACT

APOLLO 16 S-IVB IMPACT

APOLLO 13 S-IVB IMPACT

APOLLO 12

APOLLO 14 S-IVB IMPACT

OCEAN OF STORMS

SEA OF CLOUDS

GASSENDI

SEA OF MOISTURE

IMPACT POINT DISTANCE (KM)	APOLLO 12 SITE	APOLLO 14 SITE	APOLLO 15 SITE	APOLLO 16 SITE
A-12 SITE	-	181	1189	1189
A-14 SITE	181	-	1097	1008
A-15 SITE	1189	1097	-	1122
A-13 S-IVB/IU	137	-	-	-
A-14 S-IVB/IU	173	183	-	-
A-15 S-IVB/IU	355	271	1094	-
A-16 S-IVB/IU	158	304	1062	710
A-17 S-IVB/IU (TARGET)	482			

40°E. 30°E. 20°E. 10°E. 0° 10°W. 20°W. 30°W. 40°W. 50°W.
30°N. 20°N. 10°N. 0° 10°S. 20°S. 30°S.

-more-

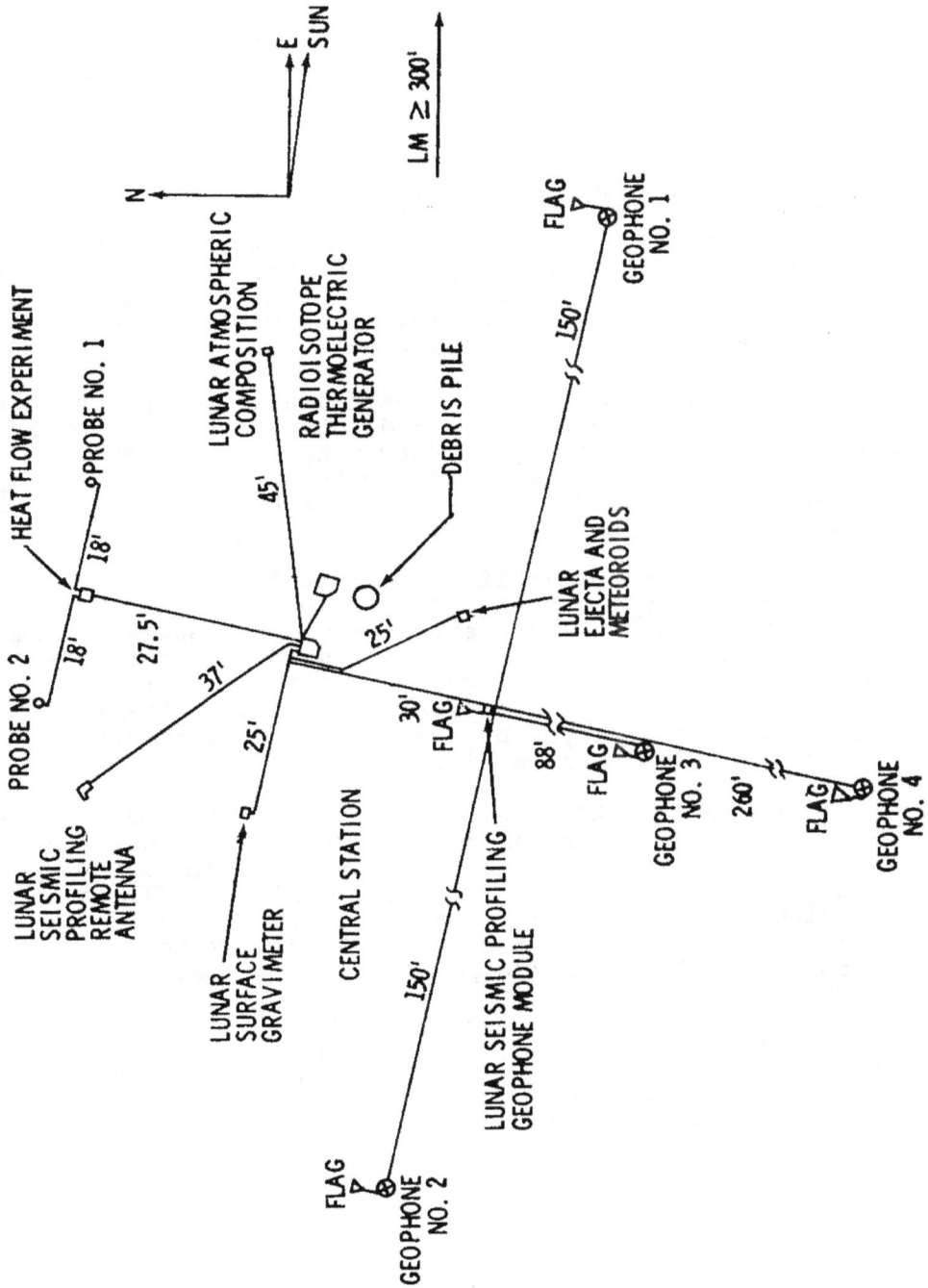

APOLLO 17
ALSEP DEPLOYMENT

SNAP-27-- Power Source for ALSEP

A SNAP-27 nuclear generator, similar to four others deployed on the Moon, will provide power for the Apollo 17 ALSEP package. The continuous operation of the four nuclear generators makes possible uninterrupted scientific surveillance of the lunar surface by the instrument packages.

SNAP-27 is one of a series of radioisotope thermoelectric generators (atomic batteries) developed by the Atomic Energy Commission under its space SNAP program. The SNAP (Systems for Nuclear Auxiliary Power) program is directed at development of generators and reactors for use in space, on land and in the sea.

SNAP-27 on Apollo 12 marked the first use of a nuclear power system on the Moon. The nuclear generators are required to provide power for at least one year. Thus far, the SNAP-27 unit on Apollo 12 has operated over three years, the unit on Apollo 14 has operated almost two years, and the unit on Apollo 15 has operated for over a year. The fourth nuclear generator was deployed by the Apollo 16 astronauts in April 1972 and is functioning normally.

The basic SNAP-27 unit is designed to produce at least 63.5 watts of electrical power. The SNAP-27 unit is a cylindrical generator, fueled with the radioisotope plutonium-238. It is about 46 cm (18 inches) high and 41 cm (16 inches) in diameter, including the heat radiating fins. The generator, making maximum use of the lightweight material beryllium, weighs about 12.7 kilograms (28 pounds) without fuel.

The fuel capsule, made of a superalloy material, is 42 cm (16.5 inches) long and 6.4 cm (2.5 inches) in diameter. It weighs about 7 km (15.5 pounds), of which 3.8 km (8.36 pounds) represent fuel. The plutonium-238 fuel is fully oxidized and is chemically and biologically inert.

The rugged fuel capsule is stowed within a graphite fuel cask from launch through lunar landing. The cask is designed to provide reentry heating protection and containment for the fuel capsule in the event of an aborted mission. The cylindrical cask with hemispherical ends includes a primary graphite heat shield, a secondary beryllium thermal shield, and a fuel capsule support structure. The cast is 58.4 cm (23 inches) long and 20 cm (eight inches) in diameter and weighs about 11 km (24.5 pounds). With the fuel capsule installed, it weighs about 18 km (40 pounds). It is mounted on the lunar module descent stage.

Once the lunar module is on the Moon, an Apollo astronaut will remove the fuel capsule from the cask and insert it into the SNAP-27 generator which will have been placed on the lunar surface near the module.

The spontaneous radioactive decay of the plutonium-238 within the fuel capsule generates heat which is converted directly into electrical energy--at least 63.5 watts. The units now on the lunar surface are producing 70 to 74 watts. There are no moving parts.

The unique properties of plutonium-238, make it an excellent isotope for use in space nuclear generators. At the end of almost 90 years, plutonium-238 is still supplying half of its original heat. In the decay process, plutonium-238 emits mainly the nuclei of helium (alpha radiation), a very mild type of radiation with a short emission range.

Before the use of the SNAP-27 system was authorized for the Apollo program a thorough review was conducted to assure the health and safety of personnel involved in the launch and of the general public. Extensive safety analyses and tests were conducted which demonstrated that the fuel would be safely contained under almost all credible accident conditions.

- more -

APOLLO LUNAR SURFACE SCIENCE MISSION ASSIGNMENTS

EXPERIMENT	11	12	14	15	16	17
S-031 PASSIVE SEISMIC	X	X	X	X	X	
S-033 ACTIVE SEISMIC			X		X	X
S-034 LUNAR SURFACE MAGNETOMETER		X		X	X	
S-035 SOLAR WIND SPECTROMETER		X		X		
S-036 SUPRATHERMAL ION DETECTOR		X	X	X		
S-037 HEAT FLOW				X	X	X
S-038 CHARGED PARTICLE LUNAR ENVIRONMENT			X			
S-058 COLD CATHODE IONIZATION		X	X	X		
M-515 LUNAR DUST DETECTOR		X	X	X		
S-207 LUNAR SURFACE GRAVIMETER						X
S-202 LUNAR EJECTA AND METEORITES						X
S-203 LUNAR SEISMIC PROFILING						X
S-205 LUNAR ATMOSPHERIC COMPOSITION						X
S-201 FAR UV CAMERA/SPECTROSCOPE					X	
S-059 LUNAR GEOLOGY INVESTIGATION	X	X	X	X	X	X
S-078 LASER RANGING RETRO-REFLECTOR	X		X	X		
S-080 SOLAR WIND COMPOSITION	X	X	X	X	X	
S-184 LUNAR SURFACE CLOSE-UP CAMERA	X	X	X			
S-152 COSMIC RAY DETECTOR						X
S-198 LUNAR PORTABLE MAGNETOMETER			X		X	
S-199 LUNAR TRAVERSE GRAVIMETER						X
S-200 SOIL MECHANICS			X	X	X	X
S-204 SURFACE ELECTRICAL PROPERTIES						X
S-229 LUNAR NEUTRON PROBE						X

-more-

Heat Flow Experiment (S-037):

The thermal conductivity and temperature gradient of the upper 2.44 meters of the lunar surface will be measured by the HFE for gathering data on the Moon's internal heating process and for a basis for comparing the radioactive content of the Moon's interior with the Earth's mantle. Results from the HFE deployed on Apollo 15 show that the Moon's outward heat flow from the interior at the Hadley-Apennine site is about half that of the Earth. The HFE principle investigator believes that the relationship between Moon and Earth heat flow is an indication that the amounts of heat-producing elements (uranium, thorium and potassium) present in lunar soil are about the same as on Earth. Temperature variations over a lunar month range at the surface from -185°C to +86°C---a spread of 271°C---but at a depth of one meter, the variations are only a few thousandths of a degree.

The Apollo Lunar Surface Drill (ALSD) is used to drill two holes about 2.6 meters deep and 10 meters apart into which the two heat flow probes are lowered. The bore stems are left in the hole and serve as casings to prevent collapse of the walls when the probes are lowered. An emplacement tool aids the crewman in ramming the probes to the maximum depth. Flat cables connect the probes to the HFE electronics package and thence to the ALSEP central station. Radiation shields are placed over each bore hole after the data cables are connected.

During deployment of the Apollo 16 HFE, the cable connecting the experiment probe to the central station was broken when an astronaut caught his foot in the cable. A fix, consisting of a strain release device, has been installed on all of the cables to preclude a reoccurrence of such an accident. It was successfully deployed on Apollo 15, however.

Principal Investigator is Dr. Marcus E. Langseth of the Lamont-Doherty Geological Observatory, Columbia University.

Lunar Ejecta and Meteorites (S-202):

The purpose of the Lunar Ejecta and Meteorites Experiment is to measure the physical parameters of primary and secondary particles impacting the lunar surface.

The detailed objectives of the experiment are: determine the background and long-term variations in cosmic dust influx rates in cislunar space; determine the extent and nature of lunar ejecta produced by meteorite impacts on the lunar surface; determine the relative contributions of comets and asteroids to the Earth's meteorite flux; study possible correlations between the associated ejecta events and times of Earth's crossing of comet orbital planes and meteor streams; determine the extent of contribution of interstellar particles toward the maintenance of the zodiacal cloud as the solar system passes through galactic space; and investigate the existence of an effect called "Earth focusing of dust particles".

The experiment package is aligned and leveled by the crew about eight meters south of the ALSEP central station using a built-in bubble level and Sun-shadow gnomon. Detector plates on the surfaces of the experiment housing are protected from lunar module ascent dust particles by a cover device which is later jettisoned by ground command.

Using a sophisticated array of detectors, the experiment measures and telemeters information such as particle velocities from 1 to 75 km/sec, energy ranges from 1 to 1000 ergs and particle impact frequencies up to 100,000 impacts per square meter per second.

Dr. Otto E. Berg, NASA Goddard Space Flight Center, is the principal investigator.

Lunar Seismic Profiling (S-203):

A major scientific tool in the exploration of the Moon has been seismology. Through this discipline, much has been learned about the structure of the lunar interior---particularly through the passive seismic experiments carried as a part of each ALSEP array from Apollo 11's initial landing through Apollo 16. The passive seismic experiments provided data on the lunar interior. All but the Apollo 11 seismometer are still operating effectively.

On Apollo 17, the experiment's data-gathering network consists of four geophones placed in the center and at each corner of a 90-meter equilateral triangle. Explosive charges placed on the surface will generate seismic waves of varying strengths to provide data on the structural profile of the landing site. The triangular arrangement of the geophones allows measurement of the azimuths and velocities of seismic waves more accurately than was possible with the Active Seismic Experiments and their linear array of geophones using mortar-fired grenades emplaced on Apollos 14 and 16.

After the charges have been fired by ground command, the experiment will settle down into a passive listening mode, detecting Moonquakes, meteorite impacts and the thump caused by the lunar module ascent stage impact. Knowledge on the surface and subsurface geologic characteristics to depths of three kilometers could be gained by the experiment. The experiment will be deactivated after the LM impact.

Components of the lunar seismic profiling experiment are four geophones similar to those used in the earlier active seismic experiment, an electronics package in the ALSEP central station, and eight explosive packages which will be deployed during the geology traverse---the lightest charge no closer than 150 meters from the geophone triangle, and the heaviest charge no further than 2.5 kilometers. Each charge has two delay timers which start after a crewman pulls three arming pins. (Timer delays vary from 90-93 hours). A coded-pulse ground command relayed through the central station will detonate each charge. Two charges weigh 57g (1/8 lb), two weigh 113g (1/4 lb), and the remaining four charges weigh 227g (1/2 lb), 454g (1 lb), 1361g (3 lb) and 2722g (6 lb) respectively. Television observations of the LSPE charge detonations as well as the LM impact are planned during the post lift-off period.

Dr. Robert L. Kovach of the Stanford University Department of Geophysics is the experiment principal investigator.

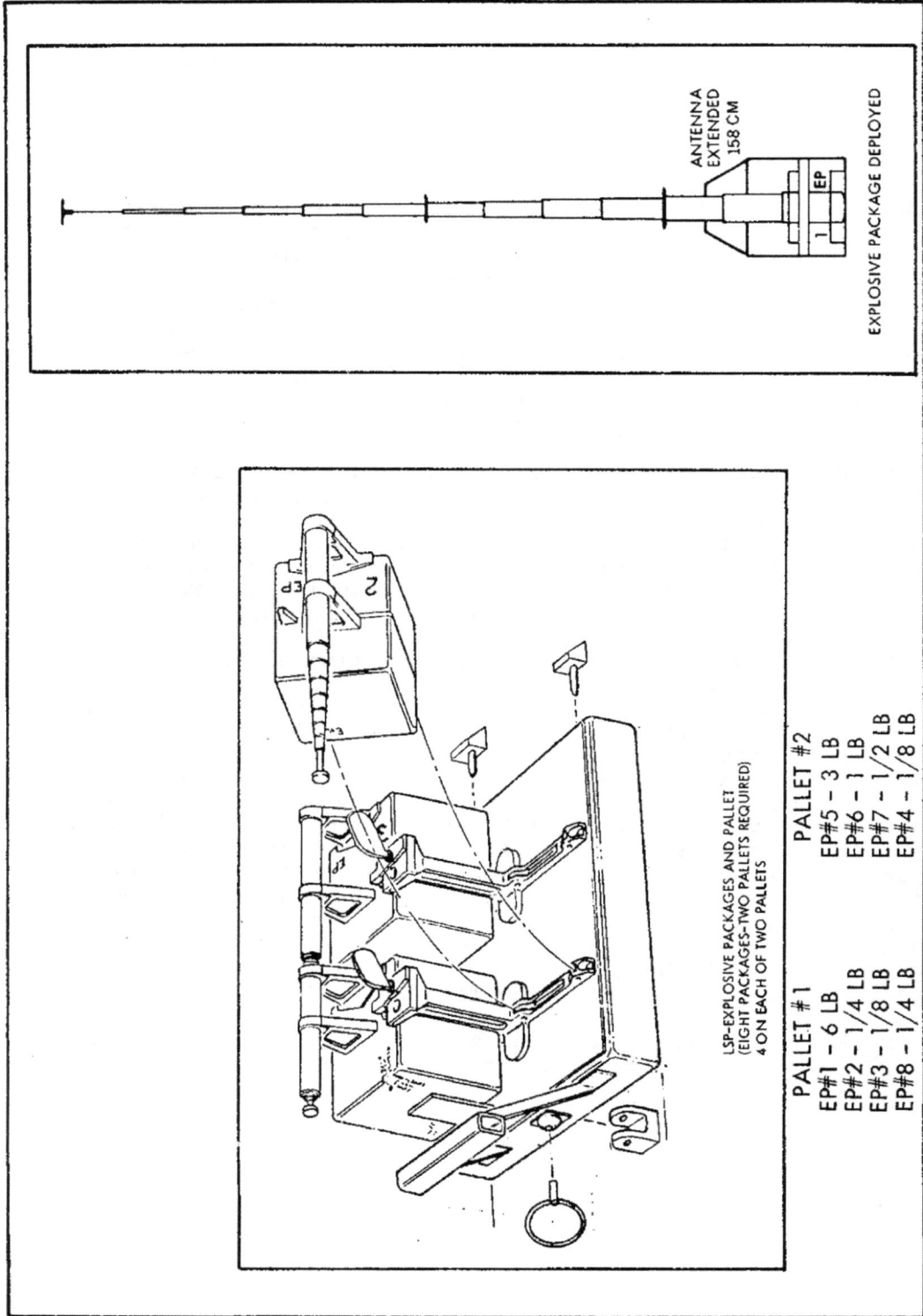

Figure 2-19(b). Lunar Seismic Profiling Experiment - Explosive Package Stowed on Pallet - Deployed Configuration

ANTENNA EXTENDED 158 CM

EXPLOSIVE PACKAGE DEPLOYED

LSP-EXPLOSIVE PACKAGES AND PALLET
(EIGHT PACKAGES-TWO PALLETS REQUIRED)
4 ON EACH OF TWO PALLETS

PALLET #1
EP#1 - 6 LB
EP#2 - 1/4 LB
EP#3 - 1/8 LB
EP#8 - 1/4 LB

PALLET #2
EP#5 - 3 LB
EP#6 - 1 LB
EP#7 - 1/2 LB
EP#4 - 1/8 LB

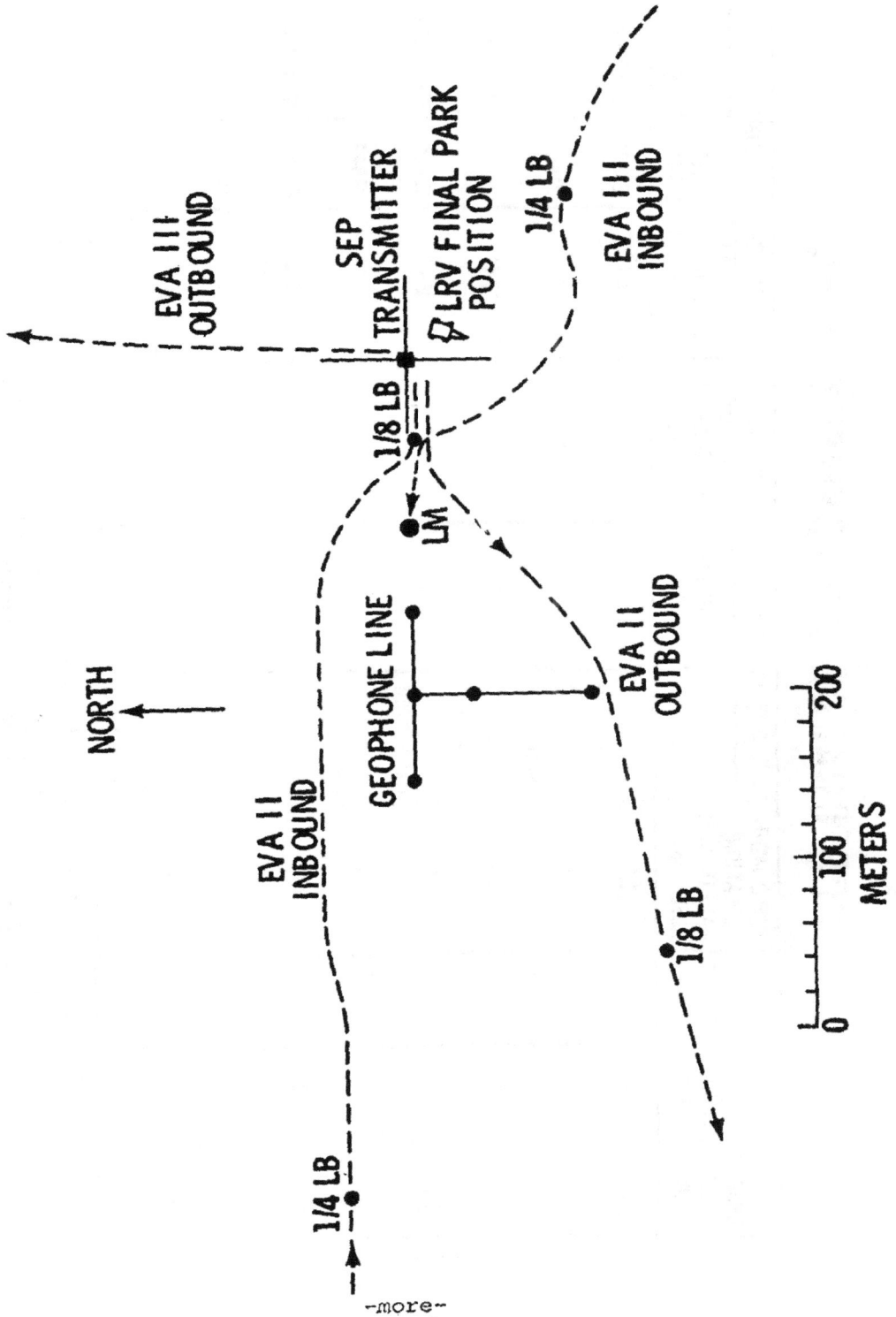

LSPE CHARGE DEPLOYMENT IN LM AREA

-42b-

-more-

NORTH

EVA III OUTBOUND

SEP TRANSMITTER

LRV FINAL PARK POSITION

1/4 LB

EVA III INBOUND

1/8 LB

LM

GEOPHONE LINE

EVA II OUTBOUND

EVA II INBOUND

1/8 LB

1/4 LB

0 100 200
METERS

LSPE CHARGE DEPLOYMENT PLAN

CHARGE SIZE LB	CHARGE NO.	PALLET NO.	DEPLOYMENT LOCATION, RADIUS FROM ALSEP (KM)	DEPLOYMENT TIME		TIME AFTER DEPLOYMENT OF DETONATION (HOURS)	TIME AFTER LIFTOFF FOR DETONATION* (HR:MIN)	TIME*	
				EVA	HR:MIN FROM EVA START			GET (HR:MIN)	DATE/EST
1	6	2	1.3	1 (OUTBOUND)	4:40	91	24:17	212:20	12/15-1813
3	5	2	2.3	1 (STATION 1)	5:53	92	26:30	214:33	12/15-2026
1/2	7	2	.8	1 (INBOUND)	6:12	93	27:49	215:52	12/15-2145
1/8	4	2	.16	2 (OUTBOUND)	:55	91	43:02	231:05	12/16-1258
6	1	1	2.4	2 (INBOUND)	5:17	92	48:24	236:27	12/16-1820
1/4	8	1	.25	2 (INBOUND)	6:11	94	51:18	239:21	12/16-2114
1/4	2	1	.25	3 (INBOUND)	5:59	93	73:36	261:39	12/17-1952
1/8	3	1	.16	3 (LRV FINAL PARKING)	6:04	94	75:07	263:10	12/17-2113

*BASED ON THE FOLLOWING MISSION TIMES:

LANDING 113:02 GET
START EVA 1 116:40 GET
START EVA 2 139:10 GET

START EVA 3 162:40 GET
LIFTOFF 188:03 GET

OTHER TIMES OF INTEREST:

LM IMPACT: 7:56 AFTER LIFT-OFF 12/15-0152 EST
TEI: 48:37 AFTER LIFT-OFF 12/16-1833 EST

Lunar Atmospheric Composition Experiment (LACE) (S-205):

This experiment will measure components in the ambient lunar atmosphere in the range of one to 110 atomic mass units (AMU). It can measure gases as thin as one billion billionth of the Earth's atmosphere. The instrument is capable of detecting changes in the atmosphere near the surface originating from the lunar module, identifying native gases and their relative mass concentrations, measuring changes in concentrations from one lunation to the next, and measuring short-term atmospheric changes.

It has been suggested that lunar volcanism may release carbon monoxide, hydrogen sulfide, ammonia, sulfur dioxide and water vapor, and that solar wind bombardment may generate an atmosphere of the noble gases - helium, neon, argon, and krypton.

The instrument can measure gases ranging from hydrogen and helium at the low end of the atomic mass scale to krypton at the high end.

The instrument is a Neir-type magnetic sector field mass spectrometer having three analyzers whose mass ranges are 1-4 AMU, 12-48 AMU and 40-110 AMU.

The experiment will be set up 15 meters northwest of ALSEP central station. After placement, it will be heated to drive off contaminating gases. The instrument will be turned on by ground command after lunar module ascent during the first lunar night and will be active thereafter.

Dr. John H. Hoffman of the Department of Atmospheric and Space Sciences of the University of Texas at Dallas is principal investigator.

Lunar Surface Gravimeter (S-207):

The major goal of the LSG is to confirm the existence of gravity waver as predicted by Einstein's general theory of relavity. Additional insights into the Moon's internal structure are expected to come from the measurement of the tidal deformation in the lunar material caused by the Earth and the Sun -- much as the Earth's ocean tides are caused by the changing Earth-Moon-Sun alignment.

Additionally the experiment is expected to detect free
Moon oscillations in periods of 15 minutes upward which may
be caused by gravitational radiation from cosmic sources.
The device will also measure vertical components of natural
lunar seismic events in frequencies up to 16 cycles per second.
It will thus supplement the passive seismic network emplaced
by Apollos 12, 14, 15 and 16.

The surface gravimeter uses a sensor based on the LaEvete-
Romberg gravimeter widely used on Earth, modified to be
remotely controlled and read. The package includes a sun-
shield, electronics and ribbon cable connecting it to the
ALSEP central station. The crew will erect the experiment
eight meters west of the central station.

Dr. Joseph Weber of the University of Maryland Department
of Physics and Astronomy is the principal investigator.

Traverse Gravimeter (S-199):

Gravimetry has proven itself to be a valuable tool for
geophysical measurements of the Earth, and the aim of the
traverse gravimeter experiment is to determine whether the
same techniques can also be used in making similar measurements of the Moon to help determine its internal structures.
Many of the major findings in geophysics, such as variations
in the lateral density in the Earth's crust and mantle, tecto-
genes, batholiths and isostasy, are a result of gravimetric
investigations.

The Apollo 17 landing site gravitational properties will
be measured in the immediate touchdown area as well as at
remote locations along the geology traverse routes. The
measurements of the Taurus-Littrow site will be related to
geologically similar areas on Earth in an effort to draw
parallels between gravimetry investigations on the two bodies.

In an extension of Earth gravimetry techniques to the
lunar surface, it is felt that small-scale lunar features
such as mare ridges, craters, rilles, scarps and thickness
variations in the regolith can better be understood.

The traverse gravimeter instrument will be mounted on
the LRV. Measurements can be made when the vehicle is not in
motion or when the instrument is placed on the lunar surface.
The crew, after starting the instrument's measurement sequence,
will read off the numbers from the digital display on the
air/ground circuit to Mission Control.

APOLLO 17

TRAVERSE GRAVIMETER

PURPOSE:

**TO MAKE A RELATIVE SURVEY
OF THE LUNAR GRAVITATIONAL FIELD
IN THE LANDING AREA AND TO MAKE
AN EARTH-MOON GRAVITY TIE**

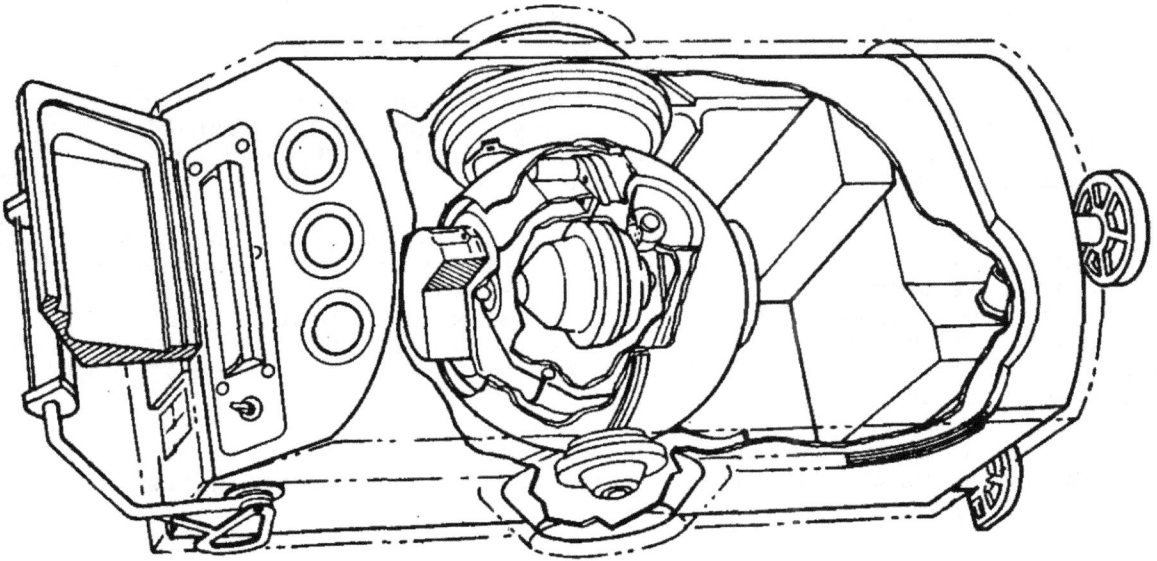

Dr. Manik Talwani of the Columbia University Lamont-Doherty Geological Observatory is the traverse gravimeter principal investigator.

Surface Electrical Properties (SEP) (S-204):

This experiment measures electromagnetic energy transmission, absorption and reflective characteristics of the lunar surface and subsurface. The instrument can measure the electrical properties at varying depths and the data gathered, when compared to the traverse gravimeter and seismic profiling data, will serve as a basis for a geological model of the upper layers of the Moon.

Frequencies transmitted downward into the lunar crust have been selected to allow determination of layering and scattering over a range of depths from a few meters to a few kilometers. Moreover, the experiment is expected to yield knowledge of the thickness of the regolith at the Taurus-Littrow site, and may provide an insight into the overall geological history of the outer few kilometers of the lunar crust.

Continuous successive waves at frequencies of 1, 2.1, 4, 8.1, 16 and 32.1 megahertz (mHz) broadcast downward into the Moon allow measurement of the size and number of scattered bodies in the subsurface. Also, any moisture present in the subsurface can easily be detected since small amounts of water in the rocks or subsoil would greatly change electrical conductivity.

The instrument consists of a deployable self-contained transmitter, having a multi-frequency antenna and a portable receiver with a wide-band three-axis antenna. The receiver contains a retrievable data recorder.

The transmitter with its antenna is deployed by the crew about 100 meters east of the lunar module, while the receiver-recorder is mounted on the LRV. The exact location of each reading is recorded on the recorder using information from the LRV's navigation system. After the final readings are made near the end of the third EVA, the recorder will be removed for return to Earth.

Dr. M. Gene Simmons of the Massachusetts Institute of Technology, Cambridge, Mass., is the principal investigator.

SURFACE ELECTRICAL PROPERTIES EXPERIMENT

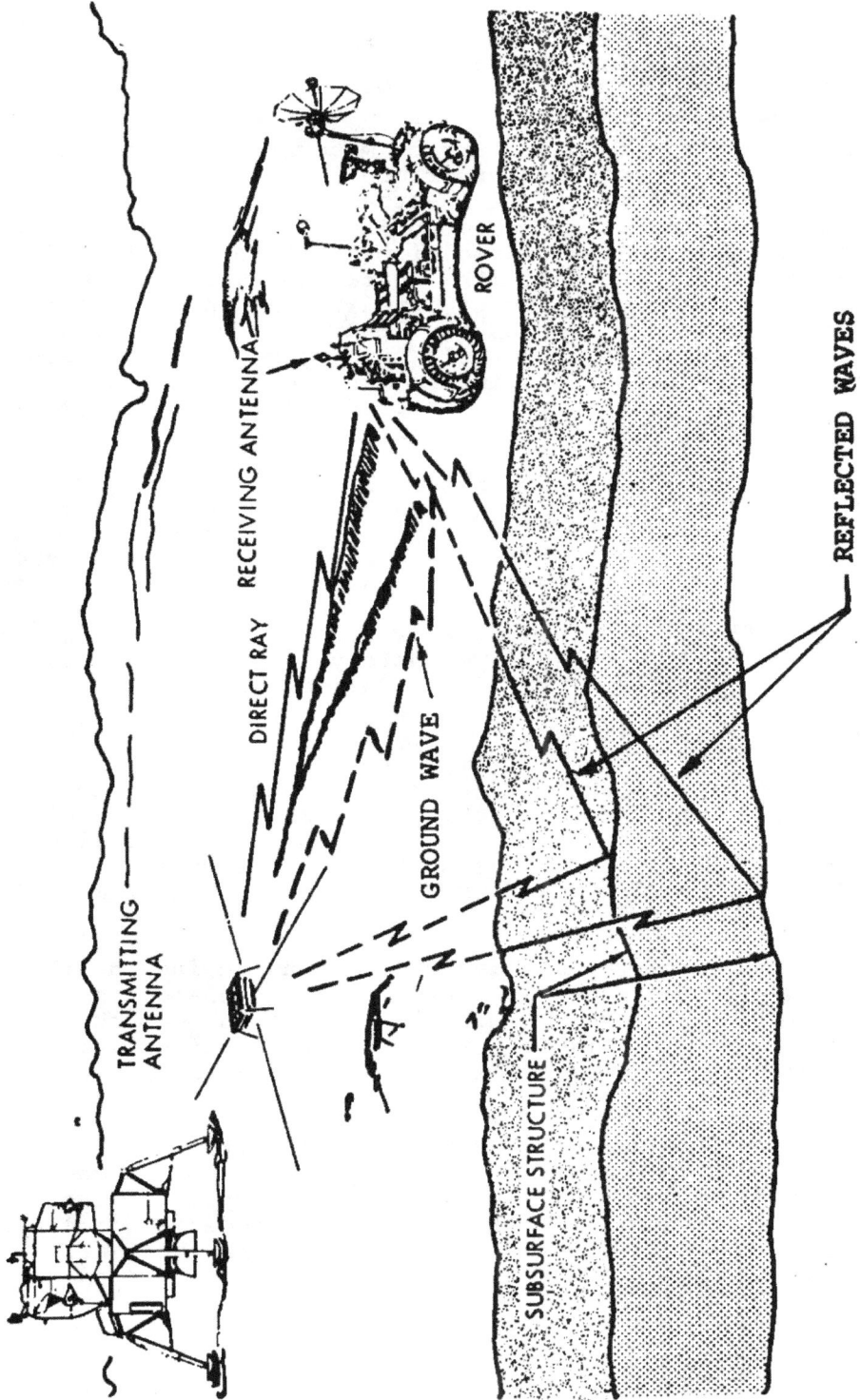

TRANSMITTING ANTENNA

RECEIVING ANTENNA

DIRECT RAY

GROUND WAVE

ROVER

REFLECTED WAVES

SUBSURFACE STRUCTURE

Lunar Neutron Probe (S-229):

Neutron capture rates of the lunar regolith and average mixing depths of lunar surface material are among the problems which it is hoped this experiment will help solve. The 2.4 meter long probe will be inserted into the hole left by the drill core sample gathered during the lunar geology experiment. It will measure the degree of present neutron flux in the top two meters of the regolith. Data from the instrument also will help determine the average irradiation depth for lunar rocks and yield information on the lunar neutron energy spectrum. The data from the probe will be compared with the distribution of gadolinium isotopes in the lunar drill core sample returned by Apollo 17 as an aid in interpreting neutron dosages on samples from previous missions.

The two-section cylindrical probe 2.4 meters long by 2 cm in diameter, is activated by the crew at the emplacement site. The crew inserts the probe in the drill core sample hole during the first EVA, and retrieves it at the end of the third EVA for deactivation and stowage for the trip home.

Principal investigator is Dr. Don S. Burnett of the California Institute of Technology, Division of Geology and Planetary Sciences, Pasadena, California.

Soil Mechanics (S-200):

While this experiment is officially listed as "passive," since no specific crew actions or hardware are involved, the wide range of knowledge gained of the physical characteristics and mechanical properties of the surface material in the landing site will make the experiment active from an information standpoint.

Crew observations and photography during the three EVAs will aid in the interpretation of lunar history and landing site geological processes, such as the form and compaction of surface layers, characteristics of rays, mares, slopes and other surface units, and deposits of different chemical and mineralogical compounds.

The experiment is further expected to contribute to knowledge on slope stability, causes of downslope movement and the natural angle of repose for different types of lunar soils. The experiment data will also aid in predicting seismic velocities in different types of material for interpreting seismic studies.

S-229 LUNAR NEUTRON PROBE

SURFACE SCIENCE EXPERIMENT

PURPOSE

MEASURE THE CAPTURE RATE
OF LOW ENERGY COSMIC RAY
SECONDARY NEUTRONS AND
NEUTRON ENERGY SPECTRUM
AS A FUNCTION OF DEPTH

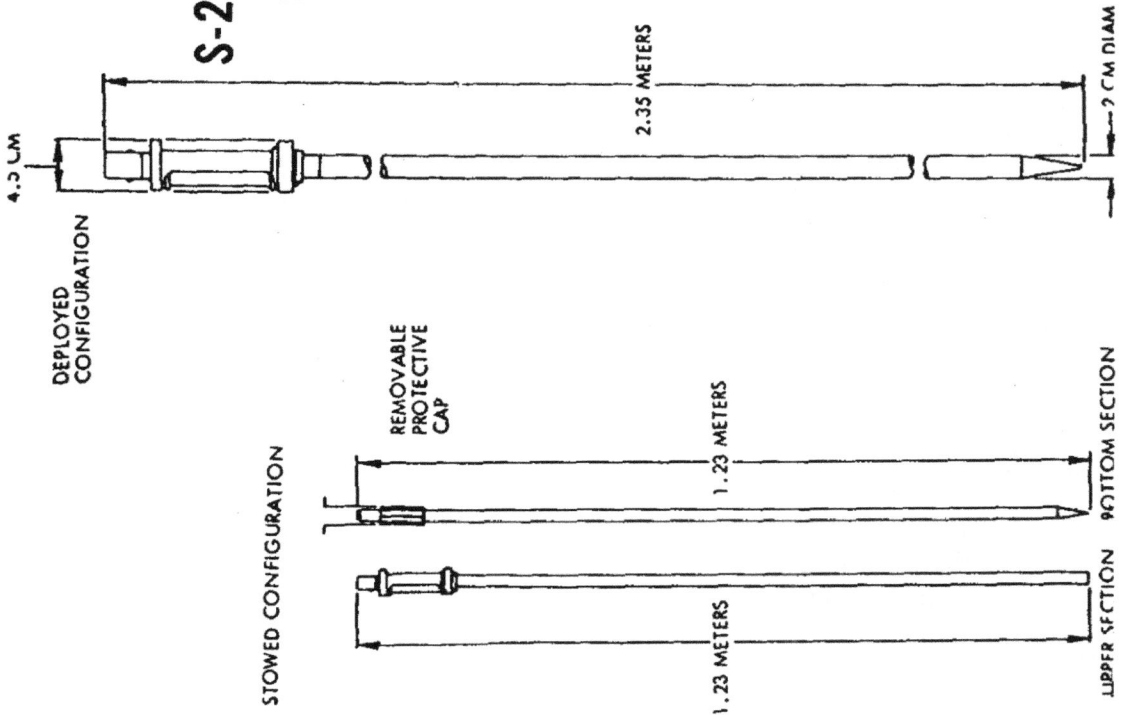

DEPLOYED
CONFIGURATION

4.5 CM

2.35 METERS

2 CM DIAM

STOWED CONFIGURATION

REMOVABLE
PROTECTIVE
CAP

1.23 METERS

1.23 METERS

BOTTOM SECTION

UPPER SECTION

Material density, specific heat and thermal conductivity data for heat-flow investigations are also expected to be supplemented by crew observations.

Among the specific items the crew will observe are the LM footpad impressions, any soil deposits on LM vertical surfaces and tracks made by the LRV during the geology traverses. These observations will aid in determining the in-place strength and compressibility of the surface material. Moreover, the observations will aid in defining conditions for simulations studies back on Earth of the returned sample density, porosity and confining pressures.

Dr. James K. Mitchell of the University of California at Berkeley is principal investigator.

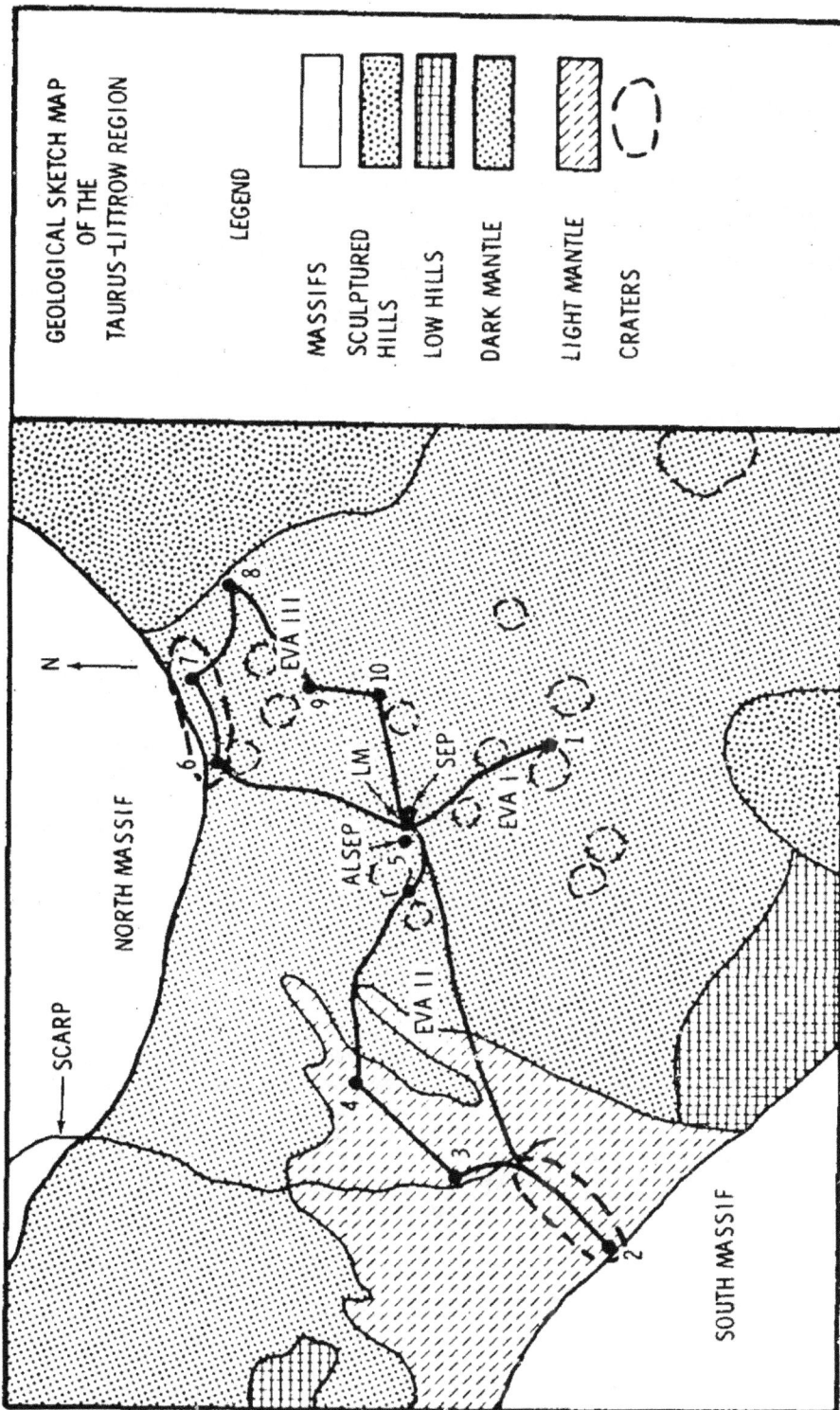

APOLLO 17
LRV TRAVERSES
(PRELIMINARY)

GEOLOGICAL SKETCH MAP
OF THE
TAURUS-LITTROW REGION

LEGEND

MASSIFS

SCULPTURED HILLS

LOW HILLS

DARK MANTLE

LIGHT MANTLE

CRATERS

SCARP

NORTH MASSIF

N

SOUTH MASSIF

Lunar Geology Investigation (S-059):

The fundamental objective of the lunar geology investigation experiment is to provide data in the vicinity of the landing site for use in the interpretation of the geologic history of the Moon. Apollo lunar landing missions offer the opportunity to correlate carefully collected samples with a variety of observational data on at least the upper portions of the mare basin filling and the lunar highlands, the two major geologic subdivisions of the Moon. The nature and origin of the maria and highlands will bear directly on the history of lunar differentiation and differentiation processes. From the lunar bedrock, structure, land forms and special materials, information will be gained about the internal processes of the Moon. The nature and origin of the debris layer (regolith) and the land forms superimposed on the maria and highland regions are a record of lunar history subsequent to their formation. This later history predominately reflects the history of the extra-lunar environment. Within and on the regolith, there will also be materials that will aid in the understanding of geologic units elsewhere on the Moon and the broader aspects of lunar history.

The primary data for the lunar geology investigation experiment come from photographs, verbal data, and returned lunar samples. Photographs taken according to specific procedures will supplement and illustrate crew comments, record details not discussed by the crew, provide a framework for debriefing, and record a wealth of lunar surface information that cannot be returned or adequately described by any other means.

In any Hasselblad picture taken from the lunar surface, as much as 90 percent of the total image information may be less than 100 feet from the camera, depending on topography and how far the camera is depressed below horizontal. Images of distant surface detail are so foreshortened that they are difficult to interpret. Therefore, it is important that panoramas be taken at intervals during the traverse and at the farthest excursion of the traverse. This procedure will extend the high resolution photographic coverage to the areas examined and discussed by the astronaut, and will show the regional context of areas of specific interest that have been discussed and photographed in detail.

- more -

The polarizing filters will permit the measurement of the degree of polarization and orientation of the plane of polarization contained in light reflected from the lunar surface. Different lunar materials (i.e., fine-grained glass and/or fragments, strongly shocked rocks, slightly shocked rocks and shock-lithified fragmental material) have different polarimetric functions, in other words, different polarimetric "signatures." Comparison of the polarimetric function of known material, such as returned samples and close-up lunar surface measurements, with materials photographed beyond the traverse of the astronaut will allow the classification and correlation of these materials even though their textures are not resolvable. The polarimetric properties of lunar materials and rock types are a useful tool for correlation and geologic mapping of each landing site, and for extrapolation of geologic data from site to site across the lunar surface.

The "in situ" photometric properties of both fine-grained materials and coarse rock fragments will serve as a basis for delineating, recognizing, describing, and classifying lunar materials. The gnomon, with photometric chart attached, will be photographed beside a representative rock and, if practical, beside any rock or fine-grained material with unusual features.

The long focal length (500 mm) lens with the HEDC will be used to provide high resolution data. A 5 to 10 centimeter resolution is anticipated at a distance of 1 to 2 km (0.6 to 1.2 mi.).

Small exploratory trenches, several centimeters deep, are to be dug to determine the character of the regolith down to these depths. The trenches should be dug in the various types of terrain and in areas where the surface characteristics of the regolith are of significant interest as determined by the astronaut crew. The main purpose of the trenches will be to determine the small scale stratigraphy (or lack of) in the upper few inches of the regolith in terms of petrological characteristics and particle size.

- more -

An organic control sample, carried in each sample return case (SRC), will be analyzed after the mission to determine the level of contamination in each SRC.

In order to more fully sample the major geological features of the Apollo 17 landing site, various groupings of sampling tasks are combined and will be accomplished in concentrated areas. This will aid in obtaining vertical as well as lateral data to be obtained in the principal geological settings. Thus, some trench samples, core tube samples and lunar environmental soil samples will be collected in association with comprehensive samples. In addition, sampling of crater rims of widely differing sizes in a concentrated area will give a sampling of the deeper stratigraphic divisions at that site. Repeating this sampling technique at successive traverse stations will show the continuity of the main units within the area.

Sampling and photographic techniques used to gather data in the landing site include:

* Documented samples of lunar surface material which, prior to gathering, are photographed in color and stereo -- using the gnomon and photometric chart for comparison of position and color properties -- to show the sample's relation to other surface features.

* Rock, boulder and soil samples of rocks from deep layers, and soil samples from the regolith in the immediate area where the rocks are gathered.

* Radial sampling of material on the rim of a fresh crater -- material that should be from the deepest strata.

* Photopanoramas for building mosaics which will allow accurate control for landing site map correlation.

* Polarimetric photography for comparison with known materials.

* Double drive tube samples to depths of 60 cm (23.6 in.) for determining the stratigraphy in multi-layer areas.

* Single drive tube samples to depths of 38 cm (15 in.) in the comprehensive sample area and in such target of opportunity areas as mounds and fillets.

* Drill core sample of the regolith which will further spell out the stratigraphy of the area sampled.

* Small exploratory trenches, ranging from 8 to 20 cm (2.4 to 9.7 in.) in depth, to determine regolith particle size and small-scale stratigraphy.

* Large equidimensional rocks ranging from 15 to 24 cm (6 to 9.4 in.) in diameter for data on the history of solar radiation. Similar sampling of rocks from 6 to 15 cm (2.4 to 6 in.) in diameter will also be made.

* Vacuum-packed lunar environment soil and rock samples kept biologically pure for postflight gas, chemical and microphysical analysis.

Dr. William R. Muehlberger of the US Geological Survey Center of Astrogeology, Flagstaff, Ariz. is the lunar geology principal investigator.

Apollo Lunar Geology Hand Tools:

Sample scale - The scale is used to weigh the loaded sample
return containers, sample bags, and other containers to
maintain the weight budget for return to Earth. The scale
has graduated markings in increments of 5 pounds to a
maximum capacity of 80 pounds. The scale is stowed and
used in the lunar module ascent stage.

Tongs - The tongs are used by the astronaut while in a
standing position to pick up lunar samples from pebble
size to fist size. The tines of the tongs are made of
stainless steel and the handle of aluminum. The tongs
are operated by squeezing the T-bar grips at the top of
the handle to open the tines. In addition to picking up
samples, the tongs are used to retrieve equipment the
astronaut may inadvertantly drop. This tool is 81 cm (32
inches) long overall.

Lunar rake - The rake is used to collect discrete samples
of rocks and rock chips ranging from 1.3 cm (one-half inch)
to 2.5 cm (one inch) in size. The rake is adjustable for
ease of sample collection and stowage. The tines, formed
in the shape of a scoop, are stainless steel. A handle,
approximately 25 cm (10 inches) long, attaches to the extension handle for sample collection tasks.

Adjustable scoop - The sampling scoop is used to collect
soil material or other lunar samples too small for the rake
or tongs to pick up. The stainless steel pan of the scoop,
which is 5 cm (2 inches) by 11 cm (41/2 inches) by 15 cm
(6 inches) has a flat bottom flanged on both sides and a
partial cover on the top to prevent loss of contents. The
pan is adjustable from horizontal to 55 degrees and 90
degrees from the horizontal for use in scooping and
trenching. The scoop handle is compatible with the extension
handle.

Hammer - This tool serves three functions; as a sampling
hammer to chip or break large rocks, as a pick, and as a
hammer to drive the drive tubes or other pieces of lunar
equipment. The head is made of impact resistant tool steel,
has a small hammer face on one end, a broad flat blade on the
other, and large hammering flats on the sides. The handle,
made of aluminum and partly coated with silicone rubber, is
36 cm (14 inches) long; its lower end fits the extension
handle when the tool is used as a hoe.

Extension handle - The extension handle extends the astronaut's reach to permit working access to the lunar surface by adding 76 cm (30 inches) of length to the handles of the scoop, rake, hammer, drive tubes, and other pieces of lunar equipment. This tool is made of aluminum alloy tubing with a malleable stainless steel cap designed to be used as an anvil surface. The lower end has a quick-disconnect mount and lock designed to resist compression, tension, torsion, or a combination of these loads. The upper end is fitted with a sliding "T" handle to facilitate any torqueing operation.

Drive Tubes - These nine tubes are designed to be driven or augured into soil, loose gravel, or soft rock such as pumice. Each is a hollow thin-walled aluminum tube 41 cm (16 inches) long and 4 cm (1.75 inch) diameter with an integral coring bit. Each tube can be attached to the extension handle to facilitate sampling. A deeper core sample can be obtained by joining tubes in series of two or three. When filled with sample, a Teflon cap is used to seal the open end of the tube, and a keeper device within the drive tube is positioned against the top of the core sample to preserve the stratigraphic integrity of the core. Three Teflon caps are packed in a cap dispenser that is approximately a 5.7 cm (2.25 inch) cube.

Gnomon and Color Patch - The gnomon is used as a photographic reference to establish local vertical Sun angle, scale, and lunar color. This tool consists of a weighted staff mounted on a tripod. It is constructed in such a way that the staff will right itself in a vertical position when the legs of the tripod are on the lunar surface. The part of the staff that extends above the tripod gimbal is painted with a gray scale from 5 to 35 percent reflectivity and a color scale of blue, orange, and green. The color patch, similarly painted in gray scale and color scale, mounted on one of the tripod legs provides a larger target for accurately determining colors in color photography.

LRV Soil Sampler - A scoop device attached to the end of the Universal Hand Tool for gathering surface soil samples and small rock fragments without dismounting from the LRV. The device has a ring 7.5 cm in diameter and a five-wire stiffening cage on the end that holds 12 telescoped plastic cap-shaped bags which are removed and sealed as each sample is taken. The sampler is 25 cm long and 7.5 cm wide.

Sample Bags - Several different types of bags are furnished
for collecting lunar surface samples. The Teflon documented
sample bag (DSB), 19 by 20 cm (7-1/2 by 8 inches) in size,
is prenumbered and packed in a 20-bag dispenser that can be
mounted on a bracket on the Hasselblad camera. Documented
sample bags (120) will be available during the lunar surface
EVAs. The sample collection bag (SCB), also 6f Teflon, has
interior pockets along one side for holding drive tubes and
exterior pockets for the special environmental sample container
and for a drive tube cap dispenser. This bag is 17 by 23 by
41 cm (6-3/4 by 9 by 16 inches) in size (exclusive of the
exterior pockets) and fits inside the sample return containers.
During the lunar surface EVAs this bag is hung on the hand or
on the portable life support system tool carrier. Four SCBs
will be carried on Apollo 17. The extra sample collection bag
(ESCB) is identical to the SCB except that the interior and
exterior pockets are omitted. During EVAs it is handled in the
same way as an SCB. Four ESCB bags will be carried on the
mission. A sample return bag, 13 by 33 by 57 cm (5 by 13 by
22.5 inches) in size, replaces the third sample return container
and is used for the samples collected on the third EVA. It
hangs on the LRV pallet during this EVA.

Long Term Surface Exposure Experiment:

Study of the abundance and composition of high energetic
particles of solar and galactic origin as well as particulate
interplanetary matter will yield important clues concerning
the chemical composition of the Sun, the formation of the
solar system, nuclear reactions in the Sun and galaxy as
well as small scale geologic surface processes on planetary
bodies.

A variety of aspects concerned with the above particles
are the subject of detailed analysis of returned lunar materials,
as well as specific instruments emplaced on the lunar surface,
e.g., Solar Wind Composition Experiment and Cosmic Ray
Detector and in particular some pieces of Surveyor III hardware
returned during the Apollo 12 mission. The lunar sample
analysis and the above instruments have yielded significant
results. However, lunar samples are exposed, on the average
too long to the space environment (in the millions of years)
which make the interpretation of the experimental results very
complex. The emplaced instruments suffer from the handicap
that they are exposed for too short a period a time (a few
hours to a few years) thus leaving some questions about the
representative character of the results. It therefore is
desirable to expose certain materials for an extended and
known time period.

This is the objective of the so called "Long Term Surface Exposure Experiment", where existing flight hardware will either be used as is or deployed in a specific fashion by the Apollo 17 crew in the hope that these materials will be retrieved at some undetermined future time, i.e., in a few decades. To investigate anticipated long-term exposure effects on the retrieved materials, precise documentation of the flight hardware has been obtained. This documentation includes detailed photography. Certain surfaces were photographed with a resolution of 80 microns. The surfaces include the LCRU-mirrors, TV-camera mirrors and lenses, Hasselblad camera lens, Heat Flow Electronics Box, LEAM Lunar Shield, and the ALSEP Helix Antenna Gimbal Housing. In addition, the chemical composition of various materials have been obtained. Most important, however, representative pieces of the above surfaces have been incorporated in the Curatorial Facilities at NASA-MSC, where they will be stored for the next few decades. These samples will serve as reference materials to those which are emplaced by the Apollo 17 crew and which hopefully will be returned.

-more-

SAMPLE RETURN BAG
(BSLSS SAMPLE BAG)

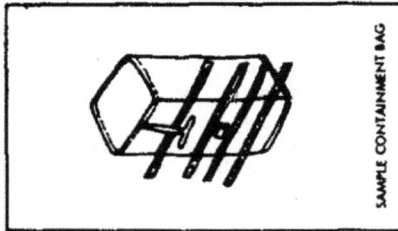

SAMPLE CONTAINMENT BAG

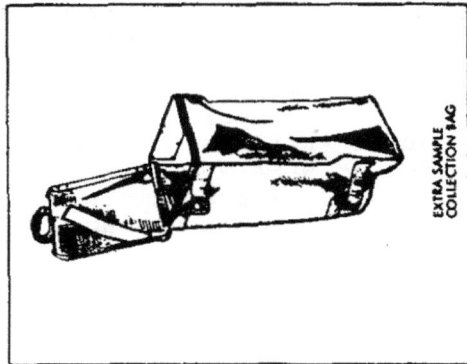

EXTRA SAMPLE
COLLECTION BAG

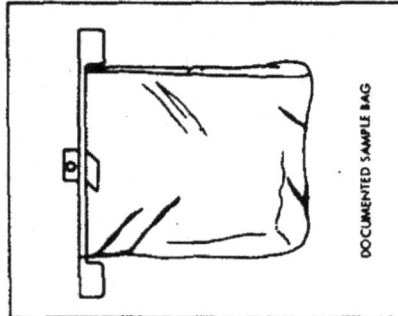

DOCUMENTED SAMPLE BAG

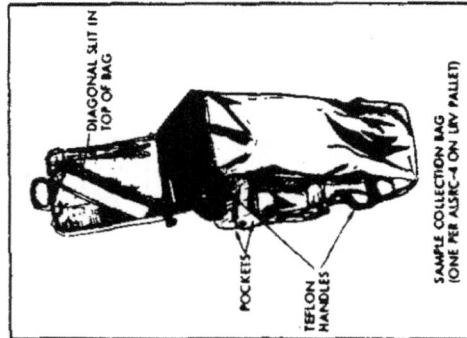

DIAGONAL SLIT IN
TOP OF BAG

SAMPLE COLLECTION BAG
(ONE PER ALSRC-4 ON LRV PALLET)

POCKETS

TEFLON
HANDLES

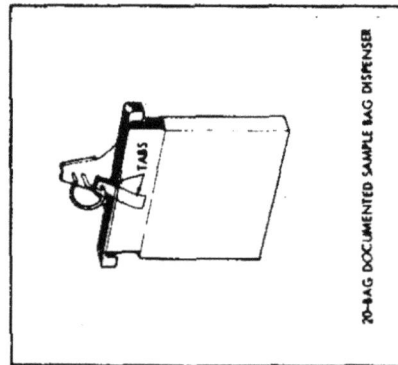

TABS

20-BAG DOCUMENTED SAMPLE BAG DISPENSER

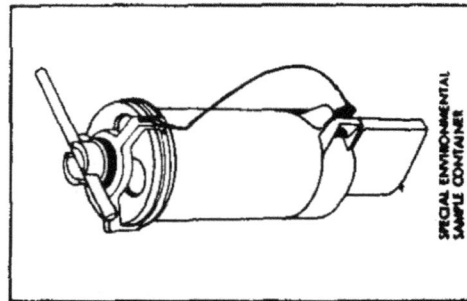

SPECIAL ENVIRONMENTAL
SAMPLE CONTAINER

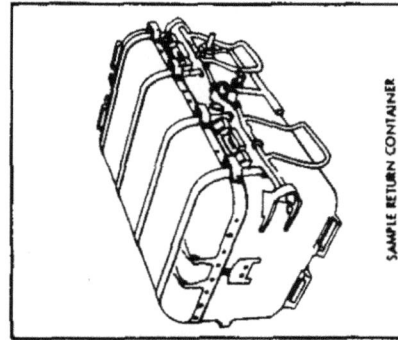

SAMPLE RETURN CONTAINER

Lunar Geology Sample Containers

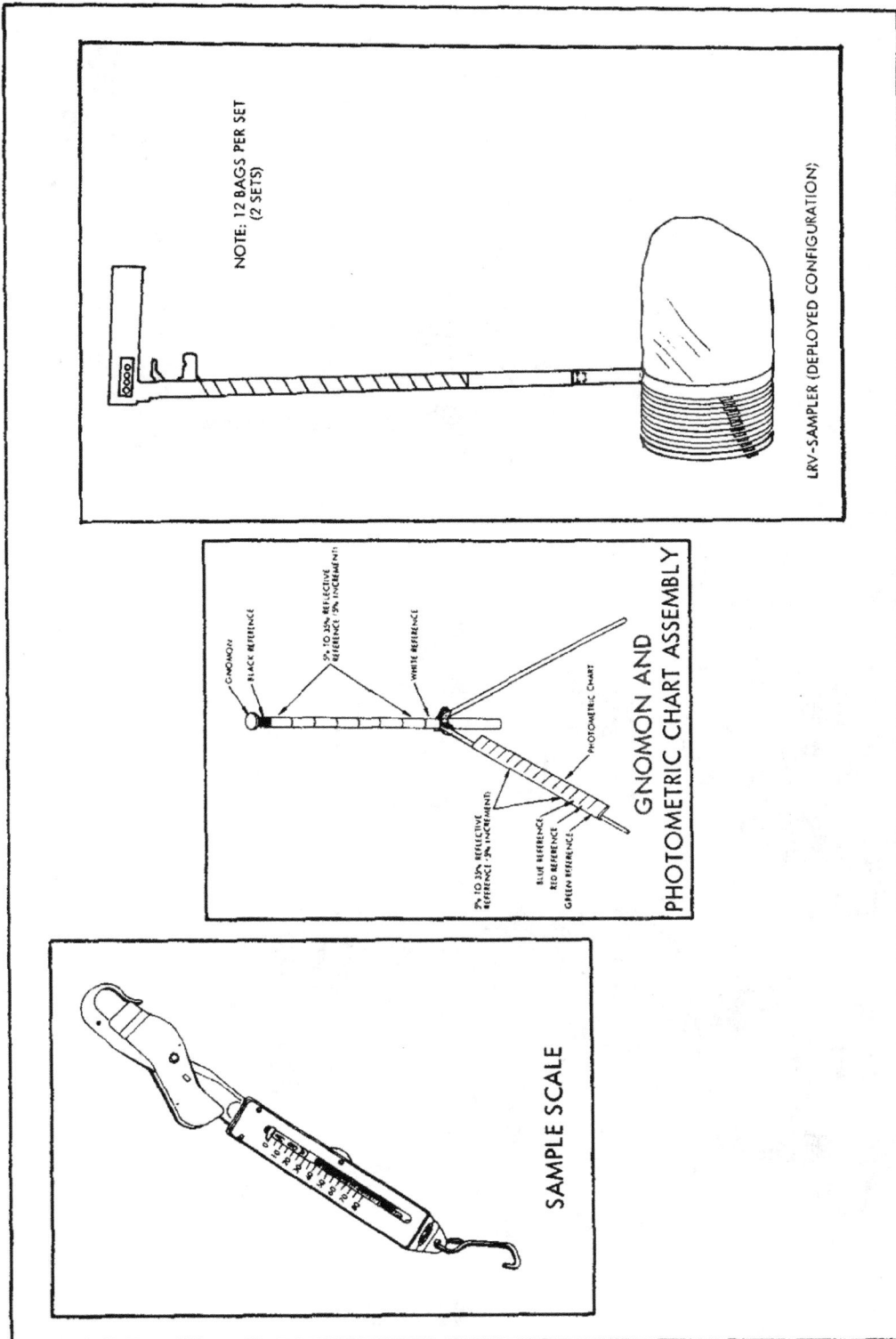

NOTE: 12 BAGS PER SET
(2 SETS)

LRV-SAMPLER (DEPLOYED CONFIGURATION)

GNOMON

BLACK REFERENCE

5% TO 35% REFLECTIVE
REFERENCE (5% INCREMENT)

WHITE REFERENCE

5% TO 35% REFLECTIVE
REFERENCE (5% INCREMENT)

PHOTOMETRIC CHART

BLUE REFERENCE
RED REFERENCE
GREEN REFERENCE

GNOMON AND
PHOTOMETRIC CHART ASSEMBLY

SAMPLE SCALE

Figure 2-22(b). Lunar Geology Equipment – Sample Scale, Gnomon/Photometric Chart, and LRV Sampler

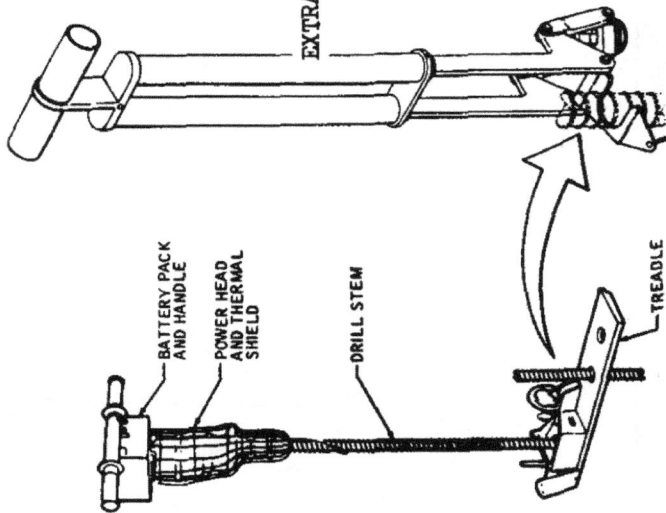

EXTRACTOR

BATTERY PACK AND HANDLE

POWER HEAD AND THERMAL SHIELD

DRILL STEM

TREADLE

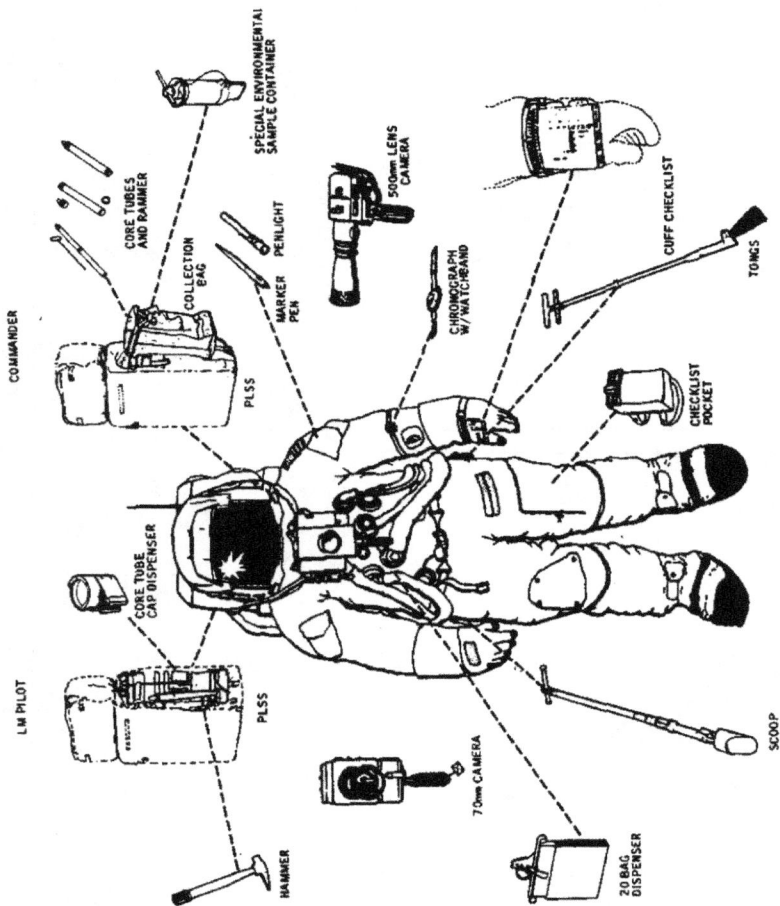

COMMANDER

CORE TUBES AND RAMMER

SPECIAL ENVIRONMENTAL SAMPLE CONTAINER

COLLECTION BAG

MARKER PEN

PENLIGHT

500mm LENS CAMERA

CHRONOGRAPH W/WATCHBAND

CUFF CHECKLIST

TONGS

PLSS

CHECKLIST POCKET

LM PILOT

CORE TUBE CAP DISPENSER

PLSS

70mm CAMERA

SCOOP

HAMMER

20 BAG DISPENSER

COMPARISON OF LUNAR SURFACE ACTIVITIES

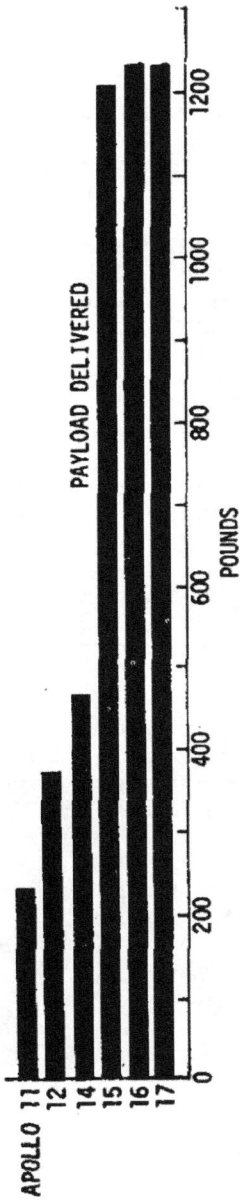

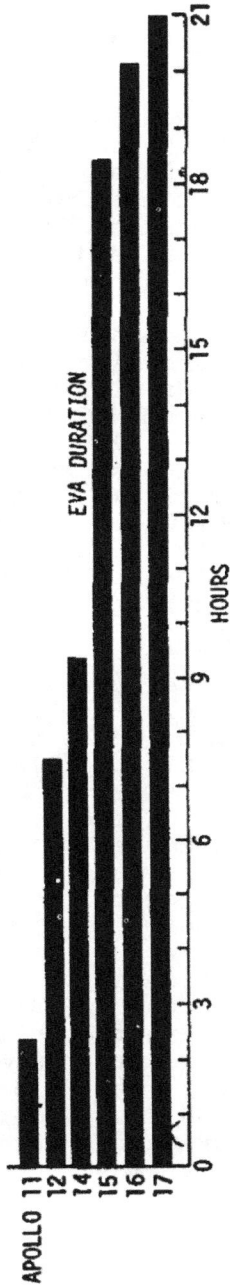

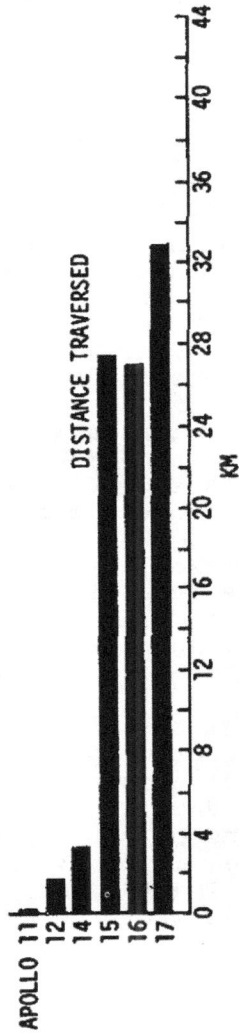

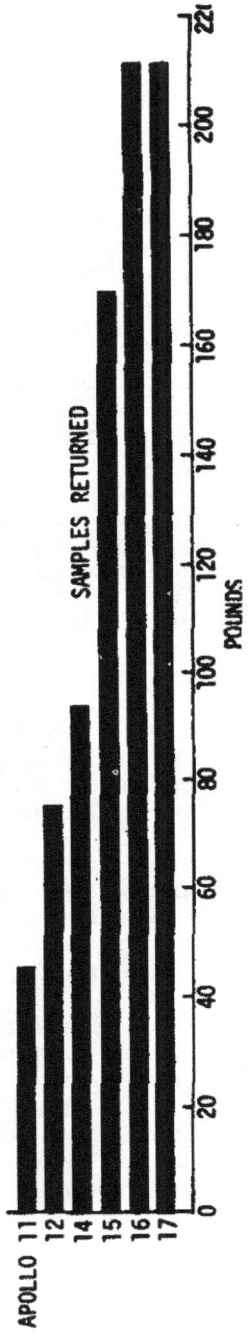

LUNAR ORBITAL SCIENCE

Service module sector 1 houses the scientific instrument module (SIM) bay. Three experiments are carried in the SIM Bay: S-209 lunar sounder, S-171 infrared scanning spectrometer and S-169 far-ultraviolet spectrometer. Also mounted in the SIM bay are the panoramic camera, mapping camera and laser altimeter used in the service module photographic tasks.

Lunar Sounder (S-209): Electromagnetic impulses beamed toward the lunar surface in the high frequency (HF) and very high frequency (VHF) bands will provide recorded data for developing a geological model of the lunar interior to a depth of 1.3 km (4,280 ft.). In addition to stratigraphic, structural, tectonic and topographic data on the region of the Moon overflown by Apollo 17, the lunar sounder will measure the ambient electromagnetic noise levels in the lunar environment at 5, 15 and 150 mHz and the occultation by the Moon of electromagnetic waves generated at the lunar surface by the surface electrical properties experiment transmitter.

Data from the lunar sounder coupled with information gathered from the SIM bay cameras and the laser altimeter, and from surface gravity measurements will allow experimenters to build an absolute topographic profile.

The experience in operating the lunar sounder and analyzing its data will contribute to the designs of future instruments for detection of surface or near-surface water on Mars, mapping of major geological units on Mars and Venus and topside sounding of Jupiter.

The lunar sounder has three major components: the coherent synthetic aperture radar (CSAR), the optical recorder, and antennas--an HF retractable dipole and a VHF yagi. The radar and optical recorder are located in the lower portions of the SIM bay. The HF dipole antenna, which has a scan of 24.4m (80 feet), deploys from the base of the SIM. The VHF yagi antenna is automatically deployed after the spacecraft/LM adapter (SLA) panels are jettisoned.

APOLLO ORBITAL SCIENCE MISSION ASSIGNMENTS

EXPERIMENT	11	12	14	15	16	17
SERVICE MODULE:						
S-160 GAMMA-RAY SPECTROMETER				X	X	
S-161 X-RAY SPECTROMETER				X	X	
S-162 ALPHA-PARTICLE SPECTROMETER				X	X	
S-164 S-BAND TRANSPONDER (CSM/LM)			X	X	X	X
S-165 MASS SPECTROMETER				X	X	
S-169 FAR UV SPECTROMETER						X
S-170 BISTATIC RADAR			X	X		
S-171 IR SCAN RADIOMETER						X
SUBSATELLITE:						
S-173 PARTICLE MEASUREMENT				X	X	
S-174 MAGNETOMETER				X	X	
S-164 S-BAND TRANSPONDER				X	X	
S-209 LUNAR SOUNDER						X
SM PHOTOGRAPHIC TASKS:						
24" PANORAMIC CAMERA				X	X	X
3" MAPPING CAMERA				X	X	X
LASER ALTIMETER				X	X	X
COMMAND MODULE:						
S-158 CM PHOTOGRAPHIC TASKS:	X		X	X	X	X
S-176 MULTISPECTRAL PHOTOGRAPHY		X				
S-177 CM WINDOW METEOROID			X	X	X	
S-177 UV PHOTO EARTH & MOON			X			
S-178 GEGENSCHEIN					X	
M-211 BIOSTACK					X	X
M-212 BIOCORE					X	X

APOLLO 17
SIM BAY

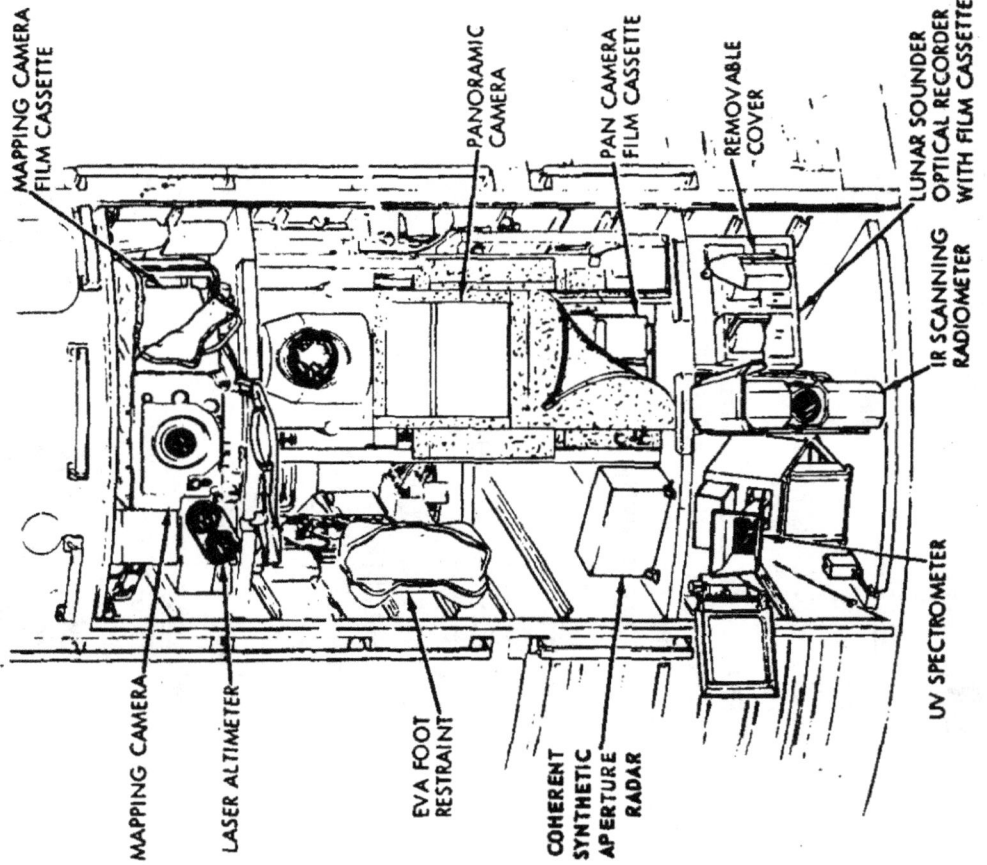

MAPPING CAMERA FILM CASSETTE

PANORAMIC CAMERA

PAN CAMERA FILM CASSETTE

REMOVABLE COVER

LUNAR SOUNDER OPTICAL RECORDER WITH FILM CASSETTE

IR SCANNING RADIOMETER

UV SPECTROMETER

MAPPING CAMERA

LASER ALTIMETER

EVA FOOT RESTRAINT

COHERENT SYNTHETIC APERTURE RADAR

APOLLO 17
LUNAR SOUNDER CONFIGURATION

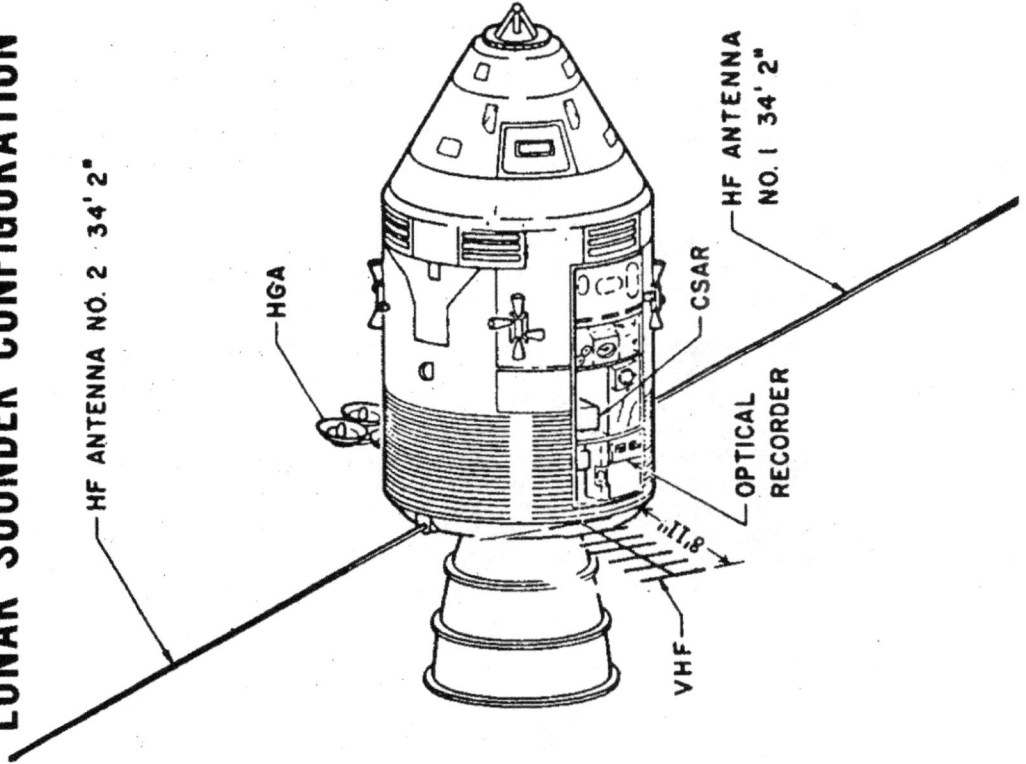

HF ANTENNA NO. 2 34' 2"

HF ANTENNA NO. I 34' 2"

HGA

CSAR

OPTICAL RECORDER

VHF

The crew has controls for deploying and jettisoning (if required) the HF antenna and for selecting lunar sounder operating modes: VHF operate, HF operate and HF receive only. During the operating modes, an electromagnetic pulse is transmitted toward the Moon and the return signal is recorded on film by the optical recorder. In the HF "receive only" mode, electromagnetic background noises from galactic sources and effects of lunar occultation are received and recorded by the instrument. During lunar sounder "operate" modes the other SIM bay instruments are powered down. The sounder's data film cassette is retrieved by the command module pilot during the trans-Earth EVA at the same time the mapping and panoramic camera cassettes are retrieved.

Dr. Stan Ward of the University of Utah and Dr. Walt Brown of the Jet Propulsion Laboratory, Pasadena, Calif., are lunar sounder principal investigators.

Infrared Scanning Radiometer (S-171): The infrared scanning radiometer (ISR) experiment will provide a lunar surface temperature map with improved temperature and spatial resolution over what has been possible before. Previous Earth-based observations of the lunar surface thermal balance have been limited to the front side with a temperature resolution of about 210°K (-80°F) and a surface resolution of about 15 km (9.3 miles). The ISR will permit measurements to be made on both the front and back side at a temperature resolution of better than 1°K (1.8°F) and a surface resolution of better than 2 km (1.2 miles). The ISR will locate and pinpoint anomalously cold or hot regions on the lunar surface in addition to measuring the surface temperature of various areas as a function of Sun angle. These cooling curves (temperature vs Sun angle) will be used to calculate the thermal properties of regions of varying geology, topography, and rock distribution.

When correlated with orbital photography and lunar sounder data, ISR temperature measurements are expected to aid in locating surface rock fields, crustal structural differences, volcanic activity and fissures emitting "hot" gases.

Mounted on the bottom shelf of the SIM Bay, the infrared scanning radiometer is made up of an optical scanning unit, a thermistor bolometer and associated processing electronics. The scanning unit consists of a folded cassegrain telescope with a rotating mirror which sweeps 162 degrees crosstrack. The thermal energy from about 2 to about 60m (7 to 200 feet) emitted from the lunar surface and reflected from the scanning mirror is focused by the telescope onto the thermistor bolometer. Changes in lunar surface temperature are detected by the bolometer. The output of the thermistor bolometer is processed by the electronics package which splits the temperature readouts into three channels for telemetry: 0-160°K (-459 to -170°F), 0-250°K (-459 to -10°F) and 0-400°K (-459 to 260°F). The lunar surface temperature at lunar noon is approximately 400°K (260°F) and drops to about 80°K (-315°F) just before lunar sunrise.

The CSM longitudinal axis will be aligned to the flight path during periods the experiment is turned on. The instrument will be operated over both the light side and darkside primarily during crew sleep periods.

Dr. Frank J. Low of the University of Arizona Lunar and Planetary Laboratory is the principal investigator.

Far-Ultraviolet Spectrometer (S-169): Atomic composition, density and scale height for several constituents of the lunar atmosphere will be measured by the far-ultraviolet spectrometer. Solar far-UV radiation reflected from the lunar surface as well as UV radiation emitted by galactic sources also will be detected by the instrument.

The far-UV spectrometer will gather data on the spectral emission in the range of 1175 to 1675 A (angstrom), and it is expected that among elements detected will be hydrogen (1216A), carbon (1675A), nitrogen (1200A), oxygen (1304A), krypton (1236A) and xenon (1470A). The total lunar atmospheric density at the lunar surface is estimated to be less than one trillionth of that at the Earth's surface.

The far-UV spectrometer, mounted on the bottom shelf of the Sim bay consists of an external baffle which limits stray light, a 0.5 meter (20 inches) focal-length Ebert mirror spectrometer with .5x6 cm (.2x2.4 inch) slits, a 10x10cm (4x4 inch) reflection grating, a scan drive mechanism which provides the wavelength scan and a photomultiplier tube which measures the intensity of the incident ultraviolet radiation, and processing electronics. The spectrometer has a 12x12 degree field of view and is aligned 18 degrees to the right of the SIM bay centerline and 23 degrees forward of the CSM vertical axis.

Controls for activating and deactivating the experiment and for opening and closing a protective cover are located in the CM.

Dr. William G. Fastie of Johns Hopkins University, Baltimore, Md., is the principal investigator.

Gamma Ray Spectrometer (S-160): As an adjunct to the gamma ray spectrometer experiment carried on Apollos 15 and 16, the aim of this task is to gather data for a calibration baseline to support the overall S-160 experiment.

A SIM bay experiment on Apollo 16, the S-160 sensor was extended on a 7.5 meter (25 foot) boom to measure natural and cosmic rays, induced gamma radioactivity from the lunar surface and the radiation flux in cislunar space. A sodium iodide crystal was used as the detector's scintillator.

An identical sodium iodide crystal will be carried aboard the Apollo 17 command module to measure background galactic radiation and CM flux reaching it during the mission. The measurements from this passive crystal will be "subtracted" from the Apollo 15 and Apollo 16 data to separate background noise from valid lunar measurements.

Immediately after splashdown, the 2.3-kilogram (five-pound) 7.5x7.5-cm (3x3-inch) cylinder containing the crystal will be removed and placed into a photomultiplier multi-channel analyzer aboard the prime recovery ship for measurements of short-period half-lives of several isotopes.

Panoramic Camera: 610mm (24-inch) SM orbital photo task:
The camera gathers mono or stereo high-resolution 2 meters (6.5ft.)
photographs of the lunar surface from orbit. The camera
produces an image size of 28 x 334 kilometers (17 x 208 nm)
with a field of view 11° along the track and 108° cross track.
The rotating lens system can be stowed face-inward to avoid
contamination during effluent dumps and thruster firings.
The 33-kilogram (72-pound) film cassette of 1,650 frames will
be retrieved by the command module pilot during a transearth
coast EVA. The camera works in conjunction with the mapping
camera and the laser altimeter to gain data to construct a
comprehensive map of the lunar surface ground track flown
by this mission ---about 2.97 million square meters (1.16
million square miles) or 8 percent of the lunar surface.

Mapping Camera: 76mm (3-inch): Combines 20-meter (66ft) resolution
terrain mapping photography on 12.5cm (5 in.) film with 76 mm
(3-inch) focal length lens with stellar camera shooting the
star field on 35mm film simultaneously at 96° from the surface
camera optical axis. The stellar photos allow accurate orientation of mapping photography postflight by comparing simulta-
neous star field photography with lunar surface photos of the
nadir (straight down). Additionally, the stellar camera
provides pointing vectors for the laser altimeter during darkside passes. The mapping camera metric lens covers a 74° square
field of view, or 170 x 170 km (92 x 92 nm) from 111.5 km (60 nm)
in altitude. The stellar camera is fitted with a 76mm (3-inch)
f/2.8 lens covering a 24° field with cone flats. The 9-kg
(20-lb) film cassette containing mapping camera film (3,600
frames) and the stellar camera film will be retrieved during
the same EVA described in the panorama camera discussion. The
Apollo Orbital Science Photographic Team is headed by
Frederick J. Doyle of the U.S. Geological Survey, McLean, Va.

Laser Altimeter: This altimeter measures spacecraft altitude
above the lunar surface to within two meters(6.5ft). The instrument
is boresighted with the mapping camera to provide altitude
correlation data for the mapping camera as well as the panoramic camera. When the mapping camera is running, the laser
altimeter automatically fires a laser pulse to the surface
corresponding to mid-frame ranging for each frame. The
laser light source is a pulsed ruby laser operating at 6,943
angstroms, and 200-millijoule pulses of 10 nanoseconds duration. The laser has a repetition rate up to 3.75 pulses per
minute. On Apollo 15 and 16, this instrument revealed important
new information on the shape of the Moon. A large depression on
the backside was found and the separation of the center of
figure and center of mass of the Moon by about 2 km (1.2 miles)
was observed. The laser altimeter working group of the Apollo
Orbital Science Photographic Team is headed by Dr. William M.
Kaula of the UCLA Institute of Geophysics and Planetary Physics.

CSM/LM S-Band Transponder: The objective of this experiment is to detect variations in lunar gravity along the
lunar surface track. These gravitational anomalies result
in minute perturbations of the spacecraft motion and are
indicative of magnitude and location of mass concentrations
on the Moon. The Spaceflight Tracking and Data Network(STDN)
and the Deep Space Network (DSN) will obtain and record
S-band doppler tracking measurements from the docked CSM/LM
and the undocked CSM while in lunar orbit; S-band doppler
tracking measurements of the LM during non-powered portions
of the lunar descent; and S-band doppler tracking measurements
of the LM ascent stage during non-powered portions of the
descent for lunar impact. The CSM and LM S-band transponders
will be operated during the experiment period. The experiment
was conducted on Apollo, 14, 15 and 16.

S-band doppler tracking data from the Lunar Orbiter missions
were analyzed and definite gravity variations were detected.
These results showed the existence of mass concentrations
(mascons) in the ringed maria. Confirmation of these results
has been obtained with Apollo tracking data, both from the CSM
and in greater detail from the Apollo 15 and 16 subsatellites.

With appropriate spacecraft orbital geometry much more
scientific information can be gathered on the lunar
gravitational field. The CSM and/or LM in low-altitude orbits
can provide new detailed information on local gravity
anomalies. These data can also be used in conjunction with
high-altitude data to possibly provide some description on
the size and shape of the perturbing masses. Correlation of
these data with photographic and other scientific records will
give a more complete picture of the lunar environment and
support future lunar activities. Inclusion of these results
is pertinent to any theory of the origin of the Moon and the
study of the lunar subsurface structure, since it implies a
rigid outer crust for the Moon for at least the last three
billion years. There is also the additional benefit of
obtaining better navigational capabilities for future lunar
missions in that an improved lunar gravity model will be known.
William Sjogren, Jet Propulsion Laboratory, Pasadena, Calif.
is principal investigator.

Apollo Window Meteoroid: This is a passive experiment in which
command module windows are scanned under high magnification
pre- and postflight for evidence of meteoroid cratering flux of
one-trillionth gram or larger. Such particle flux may be a
factor in degradation of surfaces exposed to space environment.
Principal investigator is Burton Cour-Palais, NASA Manned
Spacecraft Center.

APOLLO LUNAR ORBIT PHOTOGRAPHIC COVERAGE

	% LUNAR SURFACE		
	APOLLO 15	APOLLO 16	APOLLO 17
TOTAL AREA OVERFLOWN BETWEEN GROUNDTRACKS LOI-TEI	17.0	7.2	13.5
MAPPING CAMERA - VERTICAL			
LUNAR SURFACE PHOTOGRAPHED	10.3	5.3	8.5
NEW AREA PHOTOGRAPHED	-	3.9	2.8
TOTAL NON-REDUNDANT PHOTOGRAPHY	10.3	14.2	17.0
PAN CAMERA			
UNRECTIFIED PHOTOGRAPHY	11.5	7.2	10.3
RECTIFIED PHOTOGRAPHY	6.7	4.2	6.0
NEW AREA - RECTIFIED PHOTOGRAPHY	-	3.1	2.0
TOTAL NON-REDUNDANT RECTIFIED PHOTOGRAPHY	6.7	9.8	11.8

MEDICAL TESTS AND EXPERIMENTS

Visual Light Flash Phenomenon: Mysterious flashes of light penetrating closed eyelids have been reported by crewmen of every Apollo lunar mission since Apollo 11. Usually the light streaks and specks are observed in a darkened command module cabin while the crew is in a rest period. Averaging two flashes a minute, the phenomenon was observed in previous missions in translunar and transearth coast and in lunar orbit.

Two theories have been proposed on the origin of the flashes. One theory is that the flashes stem from visual phosphenes induced by cosmic rays. The other theory is that Cerenkov radiation by high-energy atomic particles either enter the eyeball or ionize upon collision with the retina or cerebral cortex.

The Apollo 17 crew will run a controlled experiment during translunar coast in an effort to correlate light flashes to incident primary cosmic rays. One crewman will wear an emulsion plate device on his head called the Apollo light flash moving emulsion detector (ALFMED), while his crewmates wear eyeshields. The ALFMED emulsion plates cover the front and sides of the wearer's head and will provide data on time, strength, and path of high-energy atomic particles penetrating the emulsion plates. This data will be correlated with the crewman's verbal reports on flash observations during the tests. The test will be repeated during transearth coast with all three crewman wearing eyeshields and without the ALFMED.

The experiment was also flown on Apollo 16.

Biocore (M-212): A completely passive experiment, Biocore is is designed to determine whether ionizing heavy cosmic ray particles can injure non-regenerative (nerve) cells in the eye and brain. Cosmic Ray particles range from carbon particles to iron particles or even heavier.

Pocket mice, found in the California desert near Palm Springs, are used because they are hardy, small (weighing about one-third of an ounce), and drink no water (their water comes from the seeds they eat). Five pocket mice will have cosmic ray particle detectors, in sandwich form, implanted under their scalps. These detectors are made of the plastics lexan and cellulose nitrate. Under the microscope, the tracks that cosmic ray particles make in passing through these plastics can be seen. This enables physicists to determine the path of the particles passing into the brain.

Five mice, each in a perforated aluminum tube, are housed inside an aluminum canister which is 33.8 cm long and 17.8 cm wide (12 in. long, 7 in. wide). The tubes are small so that the mice cannot float free in the zero G environment and will have ample seeds for food. A central tube in the canister contains potassium superoxide. When the animals breathe, the moisture and carbon dioxide coming from their lungs activate the superoxide which gives off oxygen to sustain the mice. The canister is a self-sustained, closed unit, not requiring any attention by the astronauts during the flight.

Principal Investigator is Dr. Webb Haymaker, NASA Ames Research Center, Mountain View, Calif.

Biostack (M-211): The German Biostack experiment is a passive experiment requiring no crew action and is quite similar to the Biostack flown on Apollo 16. The experimental results obtained on Apollo 16 were considered quite good. Conducting the experiment again on Apollo 17 with six biological materials in lieu of the four flown on Apollo 16 should enlarge the data base obtained earlier.

Selected biological material will be exposed to high-energy heavy ions in cosmic radiation and the effects analyzed postflight. Heavy ion energy measurements cannot be gathered from ground-based radiation sources. The Biostack experiments will add to the knowledge of how these heavy ions may present a hazard for man during long space flights.

Alternate layers of biological materials and radiation track detectors are hermetically sealed in an aluminum cylinder measuring 12.5 cm in diameter and 9.8 cm high (4.8 x 3.4 inches) and weighing 2.4 kg (5.3 lbs.). The cylinder will be stowed aboard the command module preflight and removed postflight for analysis by the principal investigator, Dr. Horst Bucker of the University of Frankfurt am Main, Federal Republic of Germany.

The six biological materials in Biostack, none of which is harmful to man, are bacillus subtilis spores (hay bacillis), arabiodopsis thaliana seeds (mouse-ear cress), vicia faba (broad bean roots), artemia salina eggs (brine shrimp), colpoda cucullus (protozoa cysts), and tribolium casteneum (beetle eggs).

Cardiovascular Conditioning Garment: A counterpressure garment similar to a fighter pilot's "g-suit" which will be donned by the command module pilot prior to entry and left on until completion of medical exams aboard the prime recovery vessel. Lower body negative pressure tests will be run on the CMP pre-mission and again aboard ship as part of a program to evaluate the device as a potential tool for protecting the returned crewmen in the 1G environment of Earth from cardiovascular changes to the body resulting from space flight.

Skylab Mobile Laboratories (SML) Field Test: Apollo 17 recovery operations will serve as a practical field test for the Skylab Mobile Laboratory. The SML will be loaded aboard the prime recovery vessel and staffed with 2 physicians and 4 para-medical professionals in a shakedown of the system in preparation for recovery operations for next year's Skylab mission.

The SML is made up of six basic U.S. Army Medical Unit Self-Contained Transportables (MUST) modified for Skylab postflight crew medical examinations and processing of inflight and postflight medical experiment data. Each mini-lab is outfitted to meet a specific discipline: blood, cardiovascular, metabolic studies, microbiology, nutrition and endocrinology, and operational medicine.

Food Compatibility Assessment: This investigation will measure
whole body metabolic gains or losses, together with associated
endocrinological and electrolytes controls. The purpose of
the investigation is to assess food compatibility and to determine the effect of space flight upon overall body composition
and upon the circulating and excretory levels of certain
hormonal constituents which are responsible for maintaining
homeostasis. This assessment is designed to acquire input
and output information necessary not only to assess the
metabolic consequences of the final lunar mission but also
to provide a firmer basis for interpreting the results of the
Skylab missions.

As part of this investigation, an improved urine collection system will be used by the Apollo 17 crew. Called the
Biomedical Urine Sample System (BUSS), the device consists of
a polyurethane film bag with 4,000 milliliters (120 ounces)
capacity which contains 30 milligrams (.0012 ounces) of
lithium for a tracer and 10 grams (.4 ounces) boric acid
for preservation of organic constituents. One bag is furnished
for each man/day in the command module for a total of 34.
Each bag is fitted with a roll-on cuff.

At the end of each 24-hour sampling period, a small
sample for postflight analysis is withdrawn from each of
the three BUSS bags and labeled and stowed for postflight
analysis; the remainder is vented overboard through the
command module waste management system. No samples will
be collected in lunar module operations.

ENGINEERING/OPERATIONAL
TESTS AND DEMONSTRATIONS

Heat Flow and Convection

The flow of fluids on Earth are dependent upon gravity
for motion, but in the absence of gravity in space flight
fluids behave quite differently and depend more upon surface
tension as their motive force. Investigations of fluid flow
caused by surface tension gradients, interfacial tension of
dissimilar fluids and expansion are next to impossible in
earthbound laboratories.

The Apollo 17 heat flow and convection demonstration
will go beyond the demonstration carried out on Apollo 14
by providing more exact data on the behavior of fluids in
a low gravity field -- data which will be valuable in the
design of future science experiments and for manufacturing
processes in space.

Three test cells are used in the Apollo 17 demonstration
for measuring and observing fluid flow behavior. The tests
will be recorded on motion picture film in addition to direct
observation by the crew.

Radial heat flow will be induces in a circular cell
which has an electrical heater in its center. A liquid
crystal material, which changes colors when heated, covers
the argon-filled cell, thereby indicating heat flow through
changing color patterns.

Lineal heat flow will be demonstrated in a transparent
cylinder filled with Krytox (heavy oil) and color-indicating
liquid crystal strips. Heat flow is induced by an electric
heater in one end of the cylinder.

The third device is a flow pattern test cell made up
of a shallow aluminum dish into which layers of Krytox with
suspended aluminum flakes are injected. A heater on the
bottom of the dish causes flow patterns to form in the Krytox,
with the aluminum flakes serving to make the flow more visible.

The demonstrations will be run first during translunar
coast when the spacecraft rates are nulled in all three axes,
and again after passive thermal control (PTC) or "barbecue"
roll has been set up. The 16-mm data acquisition camera
will be used to make sequence photos of the test cells -- 10
minutes for the radial and lineal test cells, and 15 minutes
for the flow pattern test cell.

Skylab Contamination Study: Since John Glenn reported seeing "fireflies" outside the tiny window of his Mercury spacecraft Friendship 7 a decade ago, space crews have noted light-scattering particles that hinder visual observations as well as photographic tasks. These clouds of particles surrounding spacecraft generally are from water dumps and escaping cabin gases changing into ice crystals.

The phenomenon could be of concern in the Skylab missions during operation of the solar astronomy experiments. The light scattering from a 100-micron particle 13 kilometers (7.8 miles) away from the spacecraft, for example, is as bright as a third-magnitude star. A cloud of particles with such a light-scattering effect would rule out any astronomical experiments being conducted on the sunlit portion of an orbit.

During translunar coast the Apollo 17 Infrared Scanning Radiometer and Far-Ultraviolet Spectrometer in the SIM Bay will provide data on optical contamination, adsorption of the contamination cloud, the scattering effect of a particle cloud, and the duration of a cloud resulting from a specific waste water dump.

Light-scattering and contamination data gathered on Apollo 17 will aid in predicting contamination in the vicinity of the Skylab orbital workshop and its sensitive telescope mount, and in devising means of minimizing contaminate levels around the Skylab vehicle. Photography in support of the study was also conducted on Apollo 16.

LUNAR ROVING VEHICLE

The lunar roving vehicle (LRV), the third to be used on the Moon, will transport two astronauts on three exploration traverses of the Moon's Taurus-Littrow region during the Apollo 17 mission. The LRV will also carry tools, scientific and communications equipment, and lunar samples.

The four- wheel, lightweight vehicle has greatly extended the lunar area that can be explored by man. It is the first manned surface transportation system designed to operate on the Moon, and it represents a solution to challenging new problems without precedent in Earth-bound vehicle design and operation.

The LRV must be folded into a small package within a wedge-shaped storage bay of the lunar module descent stage for transport to the Moon. After landing, the vehicle must be unfolded from its stowed position and deployed on the surface. It must then operate in an almost total vacuum under extremes of surface temperatures, low gravity, and on unfamiliar terrain.

The first lunar roving vehicle, used on the Apollo 15 lunar mission, was driven for three hours during its exploration traverses, covering a distance of 27.9 kilometers (17.3 statute miles) at an average speed of 9.3 kilometers an hour (5.8 miles an hour)

The second lunar roving vehicle, used on Apollo 16, was driven three hours and twenty-six minutes for a total distance of 26.9 kilometers (16.7 statute miles) at an average speed of 7.8 kilometers an hour (4.9 miles an hour).

General Description

The LRV is 3.1 meters long (10.2 feet); has a 1.8 - meter (six-foot) width; is 1.14 meters high (44.8 inches); and has a 2.3-meter wheel base (7.5 feet). Each wheel is powered by a small electric motor. The maximum speed reached on the Apollo 15 mission was 13 km/hr (eight mph), and 17 km/hr (11mph) on Apollo 16.

Two 36-volt batteries provide vehicle power, and either battery can run all systems. The front and rear wheels have separate steering systems; if one fails it can be disconnected and the LRV will operate with the other system.

Weighing about 209 kilograms (461 pounds), Earth weight, when deployed on the Moon the LRV can carry a total payload of about 490 kilograms (1,080 pounds), more than twice its own weight. The payload includes two astronauts and their portable life support systems (about 363 kilograms; 800 pounds), 68.0 kilograms (150 pounds) of communications equipment, 54.5 kilograms (150 pounds) of scientific equipment and photographic gear, and 40.8 kilograms (90 pounds) of lunar samples.

The LRV is designed to operated during a minimum period of 78 hours on the lunar surface. It can make several exploration sorties to a cumulative distance of 96 kilometers (57 miles). The maximum distance the LRV will be permitted to range from the lunar module will be approximately 9.4 kilometers (5.9 miles), the distance the crew could safely walk back to the LM in the unlikely event of a total LRV failure. This walkback distance limitation is based upon the quantity of oxygen and coolant available in the astronauts' portable life support systems. This area contains about 292 square kilometers (113 square miles) available for investigation, 10 times the area that can be explored on foot.

The vehicle can negotiate obstacles 30.5 centimeters (one foot) high and cross crevasses 70 centimeters (28 inches). The fully loaded vehicle can climb and descend slopes as steep as 25 degrees, and park on slopes up to 35 degrees. Pitch and roll stability angles are at least 45 degrees, and the turn radius is three meters (10 feet).

Both crewmen sit so the front wheels are visible during normal driving. The driver uses an on-board dead reckoning navigation system to determine direction and distance from the lunar module, and total distance traveled at any point during a traverse.

The LRV has five major systems: mobility, crew station, navigation, power, and thermal control. Secondary systems include the deployment mechanism, LM attachment equipment, and ground support equipment.

The aluminum chassis is divided into three sections that support all equipment and systems. The forward and aft sections fold over the center one for stowage in the LM. The forward section holds both batteries, part of the navigation system, and electronics gear for the traction drive and steering systems. The center section holds the crew station with its two seats, control and display console, and hand controller. The floor of beaded aluminum panels can support the weight of both astronauts standing in lunar gravity. The aft section holds the scientific payload.

Auxiliary LRV equipment includes the lunar communications relay unit (LCRU) and its high and low gain antennas for direct communications with Earth, the ground commanded television assembly (GCTA), scientific equipment, tools, and sample stowage bags.

Mobility System

The mobility system is the major LRV system, containing the wheels, traction drive, suspension, steering, and drive control electronics subsystems.

The vehicle is driven by a T-shaped hand controller located on the control and display console post between the crewmen. Using the controller, the astronaut maneuvers the LRV forward, reverse, left and right.

Each LRV wheel has a spun aluminum hub and a titanium bump stop (inner frame) inside the tire (outer frame). The tire is made of a woven mesh of zinc-coated piano wire to which titanium treads are riveted in a chevron pattern around the outer circumference. The bump stop prevents excessive inflection of the mesh tire during heavy impact. Each wheel weighs 5.4 kilograms (12 pounds) on Earth and is designed to be driven at least 180 kilometers (112 miles). The wheels are 81.3 centimeters (32 inches) in diameter and 22.9 centimeters (nine inches) wide.

A traction drive attached to each wheel has a motor harmonic drive gear unit, and a brake assembly. The harmonic drive reduces motor speed at an 80-to-1 rate for continuous operation at all speeds without gear shifting. The drive has an odometer pickup (measuring distance traveled) that sends data to the navigation system. Each motor develops 0.18 kilowatt (1/4-horsepower) and operates from a 36-volt input.

Each wheel has a mechanical brake connected to the hand controller. Moving the controller rearward de-energizes the drive motors and forces brake shoes against a drum, stopping wheel hub rotation. Full rear movement of the controller engages and locks a parking brake.

The chassis is suspended from each wheel by two parallel arms mounted on torsion bars and connected to each traction drive. Tire deflection allows a 35.6-centimeter (14-inch) ground clearance when the vehicle is fully loaded, and 43.2 centimeters (17 inches) when unloaded.

Both front and rear wheels have independent steering systems that allow a "wall-to-wall" turning radius of 3.1 meters (122 inches), exactly the vehicle length. If either set of wheels has a steering failure, its steering system can be disengaged and the traverse can continue with the active steering assembly. Each wheel can also be manually uncoupled from the traction drive and brake to allow "free wheeling" about the drive housing.

Pushing the hand controller forward increases forward speed; rear movement reduces speed. Forward and reverse are controlled by a knob on the controller's vertical stem. With the knob pushed down, the controller can only be pivoted forward; with it pushed up, the controller can be pivoted to the rear for reverse.

Crew Station

The crew station consists of the control and display console, seats, seat belts, an armrest, footrests, inboard and outboard handholds, toeholds, floor panels, and fenders.

The control and display console is separated into two main parts: The top portion holds navigation system displays; the lower portion contains monitors and controls. Attached to the upper left side of the console is an attitude indicator that shows vehicle pitch and roll.

At the console top left is a position indicator. Its outer circumference is a large dial that shows vehicle heading (direction) with respect to lunar north. Inside the dial are three digital indicators that show bearing and range to the LM and distance traveled by the LRV. In the middle of the console upper half is a Sun shadow device that is used to update the LRV's navigation system. Down the left side of the console lower half are control switches for power distribution, drive and steering, and monitors for power and temperature. A warning flag atop the console pops up if a temperature goes above limits in either battery or in any drive motor.

The LRV seats are tubular aluminum frames spanned by nylon webbing. They are folded flat during launch and erected by crewmen after deployment. The seat backs support the astronaut portable life support systems. Nylon webbing seat belts, custom fitted to each crewman, snap over the outboard handholds with metal hooks.

The armrest, located directly behind the LRV hand controller, supports the arm of the driving crewman. The footrests, attached to the center floor section, are adjusted before launch to fit each crewman. Inboard handholds help crewmen get in and out of the LRV, and have receptacles for an accessory staff and the low gain antenna of the LCRU.

The lightweight, fiberglass fenders keep lunar dust from being thrown on the astronauts, their equipment, sensitive vehicle parts, and from obstructing vision while driving. Front and rear fender sections are retracted during flight and extended by crewmen after LRV deployment on the lunar surface.

Navigation System

The navigation system is based on the principle of starting a sortie from a known point, recording speed, direction and distance traveled, and periodically calculating vehicle position.

The system has three major components: a directional gyroscope to provide vehicle headings; odometers on each wheel's traction drive unit to give speed and distance data; and a signal processing unit (a small, solid-state computer) to determine heading, bearing, range, distance traveled, and speed.

All navigation system readings are displayed on the control console. The system is reset at the beginning of each traverse by pressing a system reset button that moves all digital displays and internal registers to zero.

The directional gyroscope is aligned by measuring the inclination of the LRV (using the attitude indicator) and measuring vehicle orientation with respect to the Sun (using the shadow device).   This information is relayed to ground controllers and the gyro is adjusted to match calculated values read to back the crew.

Each LRV wheel revolution generates odometer magnetic pulses that are sent to the console displays.

Power System

The power system consists of two 36-volt, non-rechargeable batteries and equipment that controls and monitors electrical power. The batteries are in magnesium cases, use plexiglass monoblock (common cell walls) for internal construction, and have silver-zinc plates in potassium hydroxide electrolyte. Each battery has 23 cells and 121-ampere-hour capacity.

Both batteries are used simultaneously with an approximately equal load during LRV operation. Each battery can carry the entire electrical load; if one fails, its load can be switched to the other.

The batteries are activated when installed on the LRV
at the launch pad about five days before launch. During LRV
operation all mobility system power is turned off if a stop
exceeds five minutes, but navigation system power remains on
throughout each sortie. The batteries normally operate at
temperatures of 4.4 to 51.7 degrees C. (40-125 degrees F.).

An auxiliary connector at the LRV's forward end supplies
150 watts of 36-volt power for the lunar communications relay
unit.

Thermal Control

The basic concept of LRV thermal control is heat storage during vehicle operation and radiation cooling when it
is parked between sorties. Heat is stored in several thermal control units and in the batteries. Space radiators are
protected from dust during sorties by covers that are manually opened at the end of each sortie; when battery temperatures cool to about 7.2 degrees C. (45 degrees F.), the
covers automatically close.

A multi-layer insulation blanket protects forward
chassis components. Display console instruments are mounted
to an aluminum plate isolated by radiation shields and fiberglass mounts. Console external surfaces are coated with
thermal control paint and the face plate is anodized, as are
handholds, footrests, tubular seat sections, and center and
aft floor panels.

Stowage and Deployment

Space support equipment holds the folded LRV in the
lunar module during transit and deployment at three attachment points with the vehicle's aft end pointing up.

Deployment is essentially manual. One crewman releases
a cable attached to the top (aft end) of the folded LRV as
the first step in the deployment.

One of the crewmen then ascends the LM ladder part way
and pulls a D-ring on the side of the descent stage. This
releases the LRV, and lets the vehicle swing out at the top
about 12.7 centimeters (five inches) until it is stopped by
two steel cables. Descending the ladder, the crewman walks
to the LRV's right side, takes the end of a deployment tape
from a stowage bag, and pulls the tape hand-over-hand. This
unreels two support cables that swivel the vehicle outward
from the top. As the aft chassis is unfolded, the aft wheels
automatically unfold and deploy, and all latches are engaged.
The crewman continues to unwind the tape, lowering the LRV's
aft end to the surface, and the forward chassis and wheels
spring open and into place.

When the aft wheels are on the surface, the crewman removes the support cables and walks to the vehicle's left side. There he pulls a second tape that lowers the LRV's forward end to the surface and causes telescoping tubes to push the vehicle away from the LM. The two crewmen then deploy the fender extensions, set up the control and display console, unfold the seats, and deploy other equipment.

One crewman will board the LRV and make sure all controls are working. He will back the vehicle away slightly and drive it to the LM quadrant that holds the auxiliary equipment. The LRV will be powered down while the crewmen load auxiliary equipment aboard the vehicle.

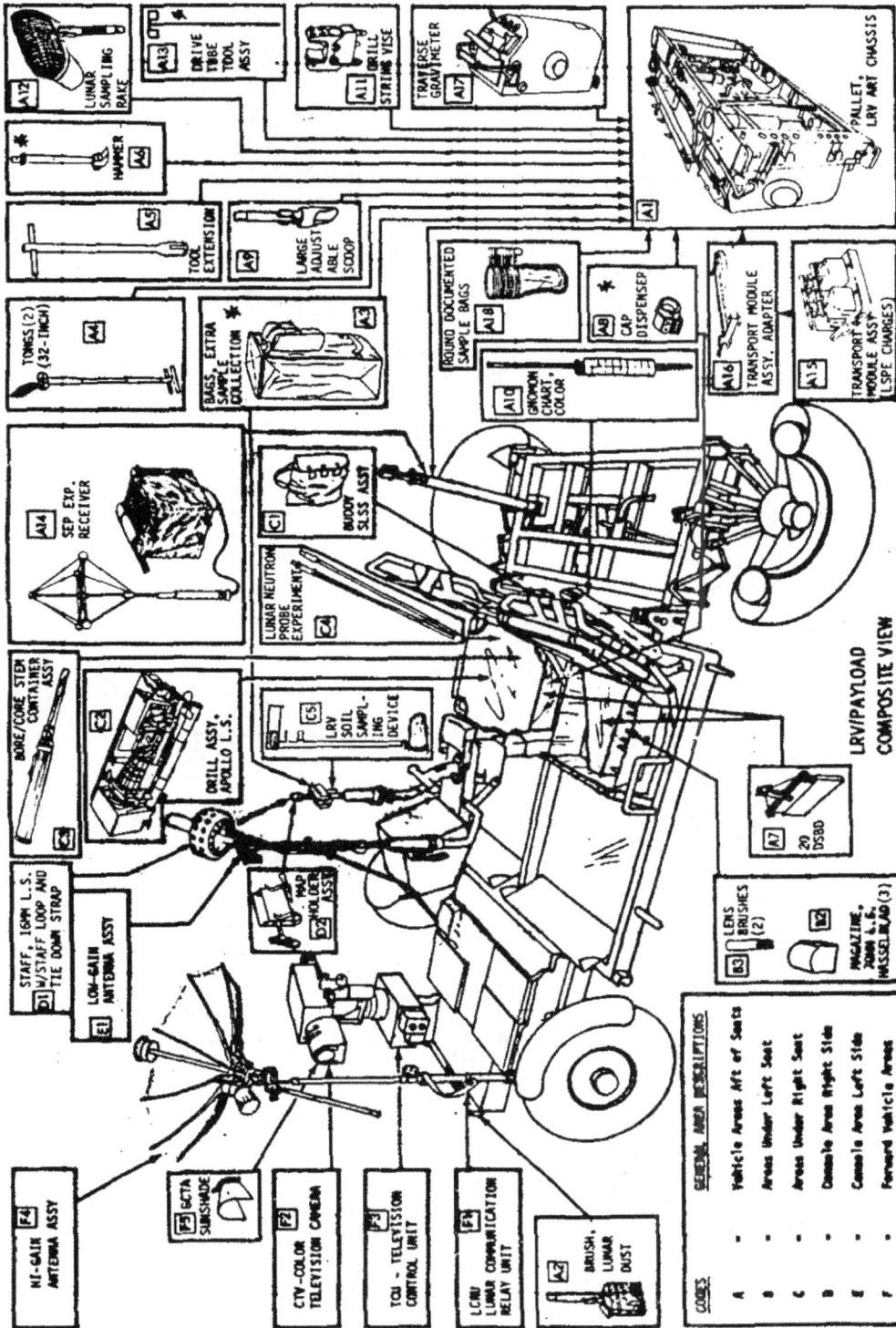

FIGURE 3.6-5 LUNAR FIELD GEOLOGY EQUIPMENT STOWAGE ON LRV

LRV DEPLOYMENT SEQUENCE

A
- LRV STOWED IN QUADRANT
- ASTRONAUT REMOVES INSULATION BLANKET, OPERATING TAPES
- ASTRONAUT REMOTELY INITIATES DEPLOYMENT

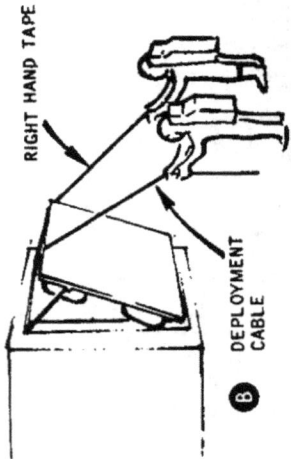

B
- RIGHT HAND TAPE
- DEPLOYMENT CABLE
- ASTRONAUT LOWERS LRV FROM STORAGE BAY WITH RIGHT HAND TAPE

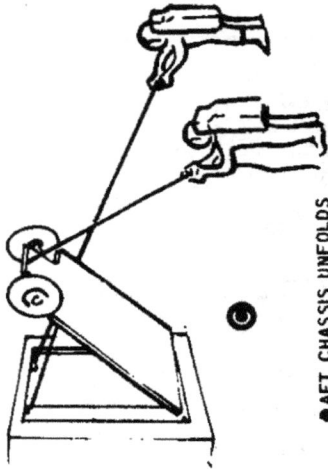

C
- AFT CHASSIS UNFOLDS
- REAR WHEELS UNFOLD
- AFT CHASSIS LOCKS IN POSITION

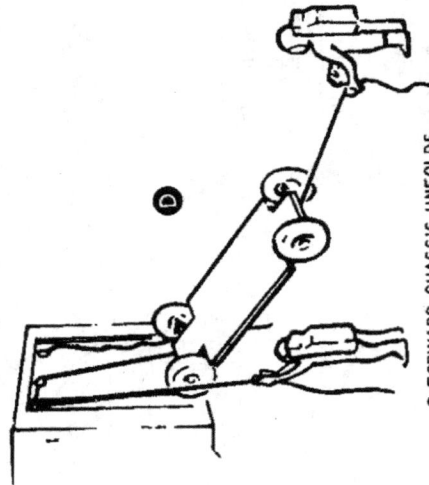

D
- FORWARD CHASSIS UNFOLDS AND LOCKS
- FRONT WHEELS UNFOLD
- ASTRONAUT LOWERS LRV TO SURFACE WITH LEFT HAND TAPE

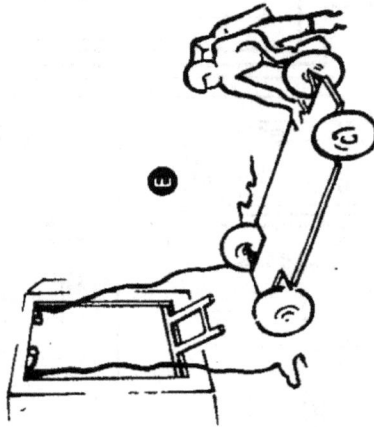

E
- ASTRONAUT DISCONNECTS SPACE SUPPORT EQUIPMENT (SSE)

HAND CONTROLLER OPERATION:

T-HANDLE PIVOT FORWARD - INCREASED DEFLECTION FROM NEUTRAL INCREASES FORWARD SPEED.

T-HANDLE PIVOT REARWARD - INCREASED DEFLECTION FROM NEUTRAL INCREASES REVERSE SPEED.

T-HANDLE PIVOT LEFT - INCREASED DEFLECTION FROM NEUTRAL INCREASES LEFT STEERING ANGLE.

T-HANDLE PIVOT RIGHT - INCREASED DEFLECTION FROM NEUTRAL INCREASES RIGHT STEERING ANGLE.

T-HANDLE DISPLACED REARWARD - REARWARD MOVEMENT INCREASES BRAKING FORCE. FULL 3 INCH REARWARD APPLIES PARKING BRAKE. MOVING INTO BRAKE POSITION DISABLES THROTTLE CONTROL AT 15° MOVEMENT REARWARD.

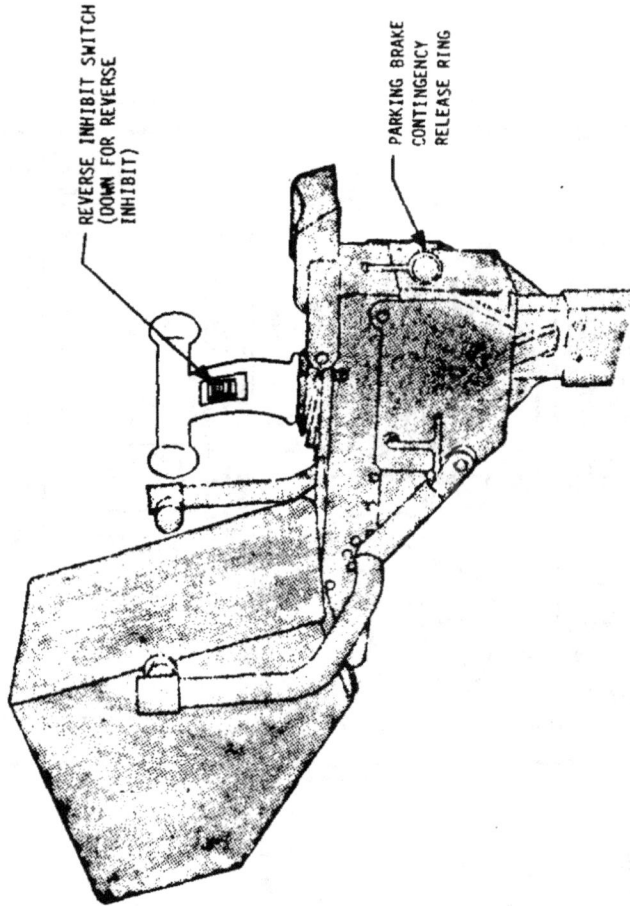

REVERSE INHIBIT SWITCH (DOWN FOR REVERSE INHIBIT)

PARKING BRAKE CONTINGENCY RELEASE RING

HAND CONTROLLER

-more-

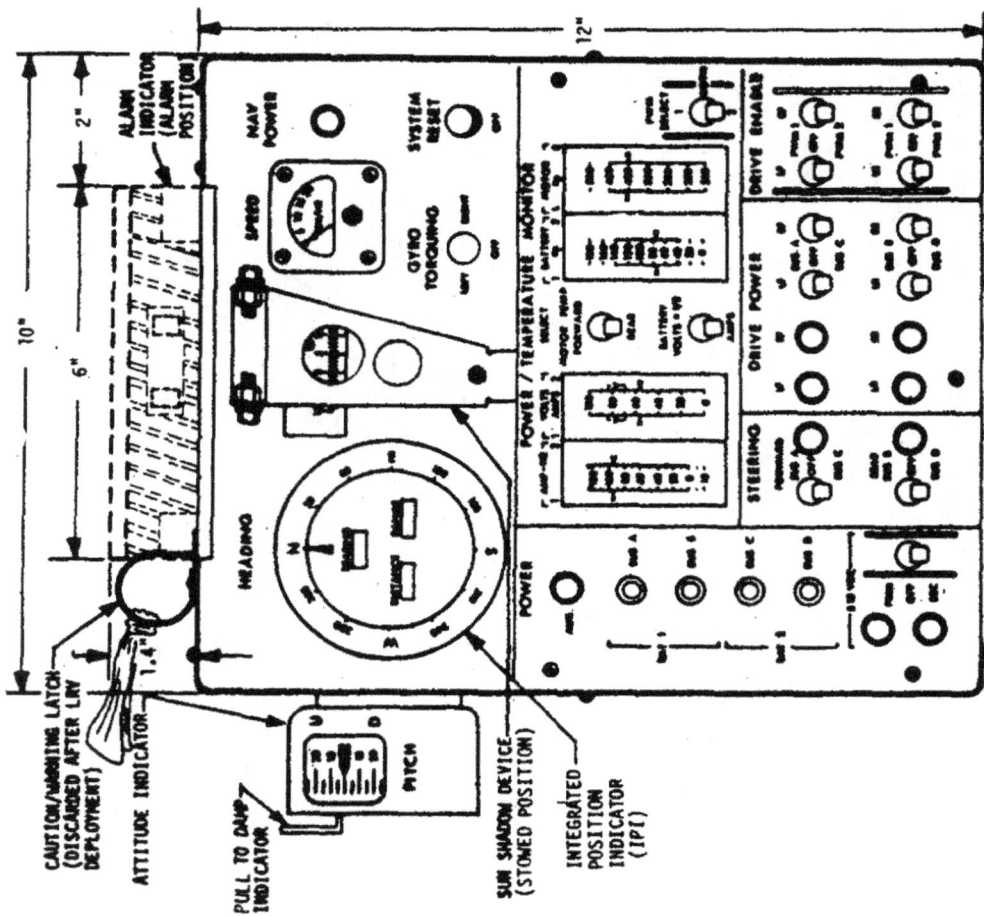

LRV CREW STATION COMPONENTS - CONTROL AND DISPLAY CONSOLE

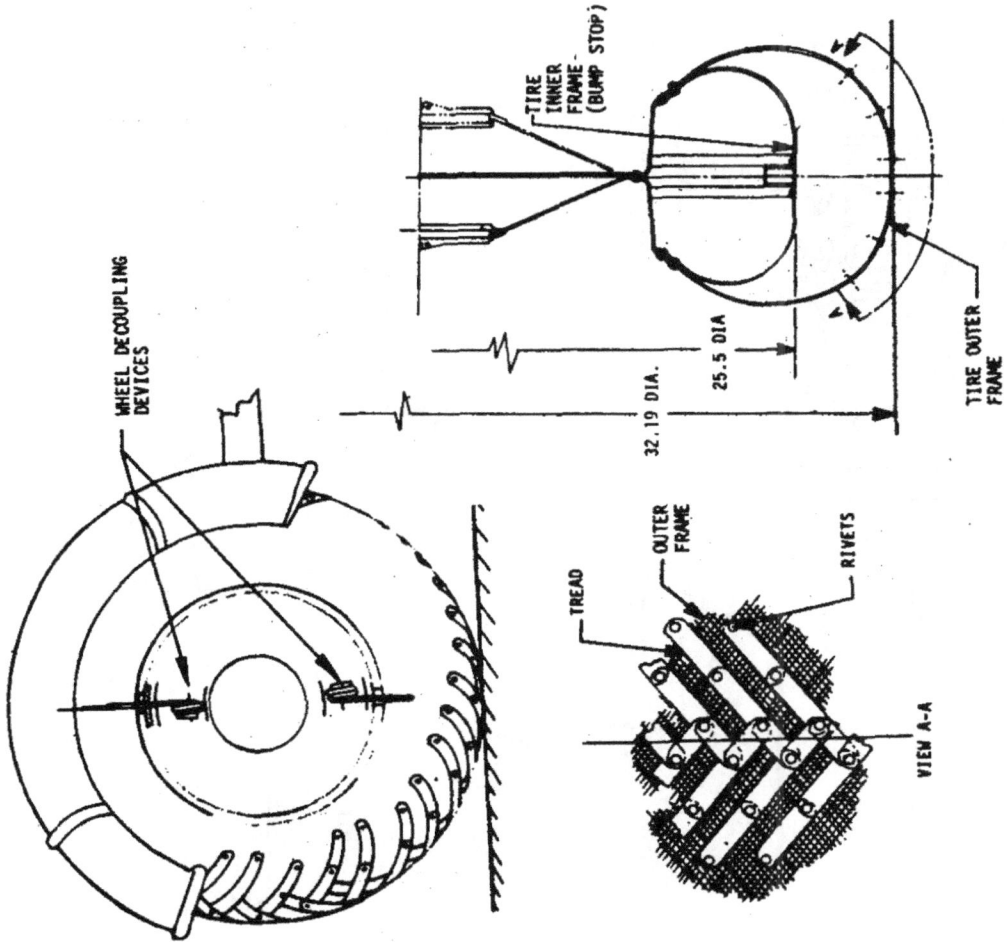

WHEEL DECOUPLING
DEVICES

TIRE
INNER
FRAME
(BUMP STOP)

25.5 DIA

32.19 DIA.

TIRE OUTER
FRAME

LRV WHEEL

OUTER
FRAME

TREAD

RIVETS

VIEW A-A

AXIS REFERENCE

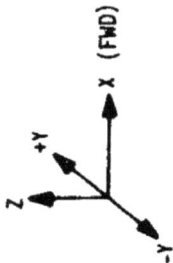

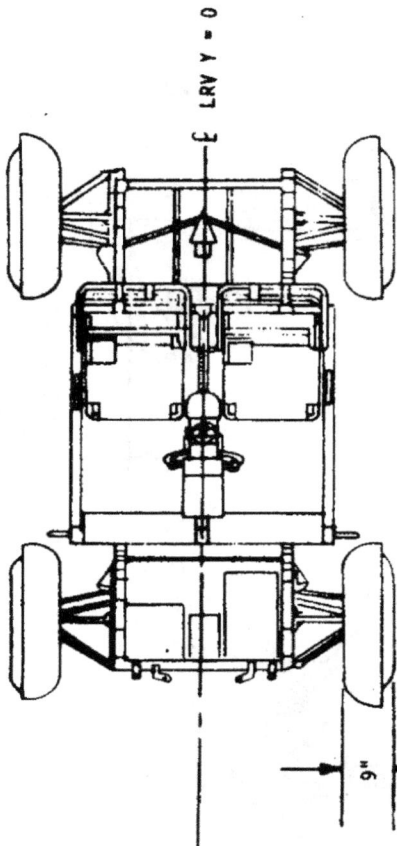

X (FWD)

Z

+Y

-Y

(DEPLOYED, EMPTY)

WEIGHT = 462 LB*

C.G. LOCATION:

X = 52.8
Y = -0.3
Z = 103.1

*INCLUDES BATTERIES
& PAYLOAD SUPPORTS,
EXCLUDES SSE.

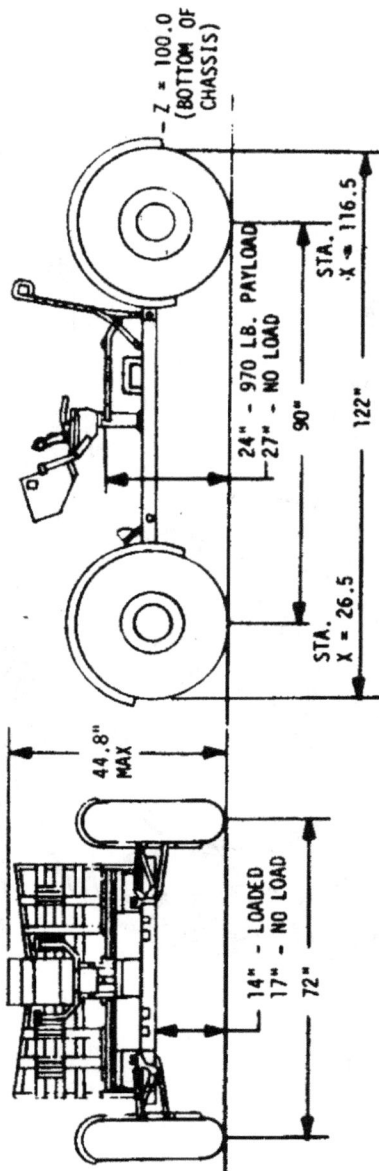

¢ LRV Y = 0

9"

Z = 100.0
(BOTTOM OF
CHASSIS)

STA.
X = 116.5

24" - 970 LB. PAYLOAD
27" - NO LOAD

90"

122"

STA.
X = 26.5

44.8"
MAX

14" - LOADED
17" - NO LOAD

72"

LRV COMPONENTS AND DIMENSIONS

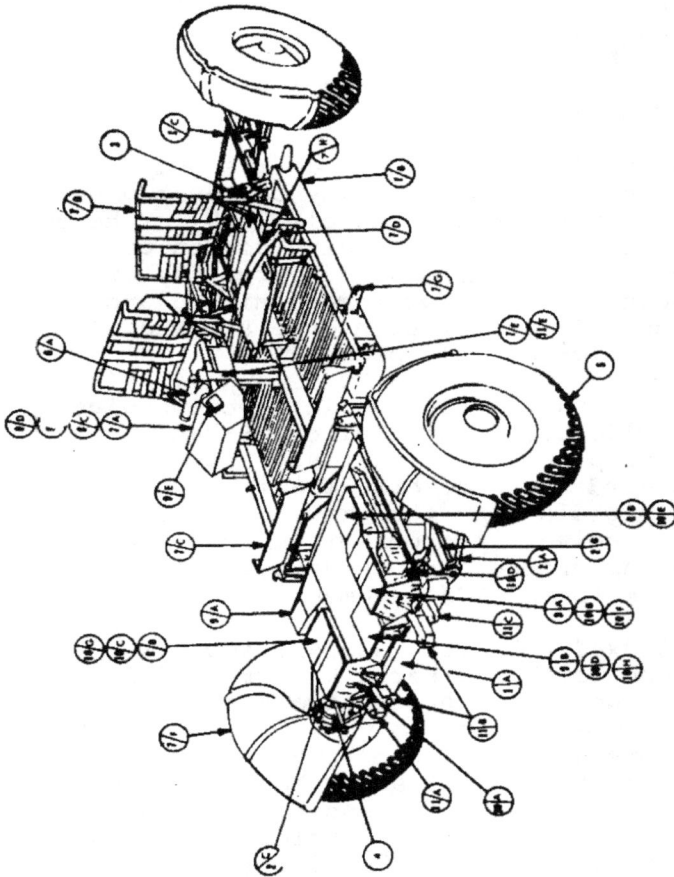

1 CHASSIS
A. FORWARD CHASSIS
B. CENTER CHASSIS
C. AFT CHASSIS

2 SUSPENSION SYSTEM
A. SUSPENSION ARMS (UPPER AND LOWER)
B. TORSION BARS (UPPER AND LOWER)
C. DAMPER

3 STEERING SYSTEM (FORWARD AND AFT)
4 TRACTION DRIVE
5 WHEEL
6 DRIVE CONTROL
A. HAND CONTROLLER
B. DRIVE CONTROL ELECTRONICS (DCE)

7 CREW STATION
A. CONTROL AND DISPLAY CONSOLE
B. SEAT
C. FOOTREST
D. OUTBOARD HANDHOLD
E. INBOARD HANDHOLD
F. FENDER
G. TOEHOLD
H. SEAT BELT

8 POWER SYSTEM
A. BATTERY #1
B. BATTERY #2
C. INSTRUMENTATION

9 NAVIGATION
A. DIRECTIONAL GYRO UNIT (DGU)
B. SIGNAL PROCESSING UNIT (SPU)
C. INTEGRATED POSITION INDICATOR (IPI)
D. SUN SHADOW DEVICE
E. VEHICLE ATTITUDE INDICATOR

10 THERMAL CONTROL
A. INSULATION BLANKET
B. BATTERY NO. 1 DUST COVER
C. BATTERY NO. 2 DUST COVER
D. SPU DUST COVER
E. DCE THERMAL CONTROL UNIT
F. BATTERY NO. 1 RADIATOR
G. BATTERY NO. 2 RADIATOR
H. SPU THERMAL CONTROL UNIT

11 PAYLOAD INTERFACE
A. TV CAMERA RECEPTACLE
B. LCRU RECEPTACLE
C. HIGH GAIN ANTENNA RECEPTACLE
D. AUXILIARY CONNECTOR
E. LOW GAIN ANTENNA RECEPTACLE

LRV WITHOUT STOWED PAYLOAD

LUNAR COMMUNICATIONS RELAY UNIT (LCRU)

The range from which an Apollo crew can operate from the lunar module during EVAs while maintaining contact with the Earth is extended over the lunar horizon by a suitcase-size device called the lunar communications relay unit (LCRU). The LCRU acts as a portable direct relay station for voice, TV, and telemetry between the crew and Mission Control Center instead of through the lunar module communications system. First use of the LCRU was on Apollo 15.

Completely self-contained with its own power supply and erectable hi-gain S-Band antenna, the LCRU may be mounted on a rack at the front of the lunar roving vehicle (LRV) or hand-carried by a crewman. In addition to providing communications relay, the LCRU receives ground-command signals for the ground commanded television assembly (GCTA) for remote aiming and focusing the lunar surface color television camera. The GCTA is described in another section of this press kit.

Between stops with the lunar roving vehicle, crew voice is beamed Earthward by a wide beam-width helical S-Band antenna. At each traverse stop, the crew must sight the high-gain parabolic antenna toward Earth before television signals can be transmitted. VHF signals from the crew portable life support system (PLSS) transceivers are converted to S-Band by the LCRU for relay to the ground, and conversely, from S-Band to VHF on the uplink to the EVA crewmen.

The LCRU measures 55.9x40.6x15.2cm (22x16x6 inches) not including antennas, and weighs 25 Earth kilograms (55 Earth pounds) (9.2 lunar pounds). A protective thermal blanket around the LCRU can be peeled back to vary the amount of radiation surface which consists of 1.26 m^2 (196 square inches) of radiating mirrors to reflect solar heat. Additionally, wax packages on top of the LCRU enclosure stabilize the LCRU temperature by a melt-freeze cycle. The LCRU interior is pressurized to 7.5 psia differential (one-half atmosphere).

Internal power is provided to the LCRU by a 19-cell silver-zinc battery with a postassium hydroxide electrolyte. The battery weighs 4.1 kg (nine Earth lbs.) (1.5 lunar lbs.) and measures 11.8x23.9x11.8cm (4.7x9.4x4.65 inches). The battery is rated at 400 watt hours, and delivers 29 volts at a 3.1-ampere current load. The LCRU may also be operated from the LRV batteries. The nominal plan is to operate the LCRU using LRV battery power during EVA-1. The LCRU battery will provide the power during EVA-2 and EVA-3.

Three types of antennas are fitted to the LCRU system: a low-gain helical antenna for relaying voice and data when the LRV is moving and in other instances when the high-gain antenna is not deployed; a .9 m (three-foot) diameter parabolic rib-mesh high-gain antenna for relaying a television signal; and a VHF omni-antenna for receiving crew voice and data from the PLSS transceivers. The high-gain antenna has an optical sight which allows the crewman to boresight on Earth for optimum signal strength. The Earth subtends a two degree angle when viewed from the lunar surface.

The LCRU can operate in several modes: mobile on the LRV, fixed base such as when the LRV is parked, or hand-carried in contingency situations such as LRV failure. The LCRU is manufactured by RCA.

TELEVISION AND GROUND
COMMANDED TELEVISION ASSEMBLY

Two different color television cameras will be used during the Apollo 17 mission. One, manufactured by Westinghouse, will be used in the command module. It will be fitted with a 5-centimeter (2-inch) black and white monitor to aid the crew in focus and exposure adjustment.

The other camera, manufactured by RCA, is for lunar surface use and will be operated from the lunar roving vehicle (LRV) with signal transmission through the lunar communication relay unit rather than through the LM communications system.

While on the LRV, the camera will be mounted on the ground commanded television assembly (GCTA). The camera can be aimed and controlled by astronauts or it can be remotely controlled by personnel located in the Mission Control Center. Remote command capability includes camera "on" and "off", pan, tilt, zoom, iris open/closed (f2.2 to f22) and peak or average automatic light control.

The GCTA is capable of tilting the TV camera upward 85 degrees, downward 45 degrees, and panning the camera 350 degrees between mechanical stops. Pan and tilt rates are approximately 3 degrees per second.

The TV lens can be zoomed from a focal length of 12.5mm to 75mm corresponding to a field of view from 9 to 54 degrees.

At the end of the third EVA, the crew will park the LRV about 91.4 m (300 ft.) east of the LM so that the color TV camera can cover the LM ascent from the lunar surface. Because of a time delay in a signal going the quarter million miles out to the Moon, Mission Control must anticipate ascent engine ignition by about two seconds with the tilt command.

The GCTA and camera each weigh approximately 5.9 kg (13 lb.). The overall length of the camera is 46 cm (18.0 in.) its width is 17 cm (6.7 in.), and its height is 25 cm (10 in.). It is powered from the LCRU battery supply, or externally from the LRV batteries. The GCTA is built by RCA.

APOLLO 17
TELEVISION EVENTS

DATE	TIME (GET)	TIME (EST)	DURATION (HRS:MIN)	EVENT
7 DEC	4:12	0205	0:20	TD & E
11 DEC	117:55	1948	5:19	EVA-1
12 DEC	139:38	1731	6:21	EVA-2
13 DEC	163:05	1658	6:35	EVA-3
14 DEC	187:48	1741	0:25	LM LIFT-OFF
14 DEC	189:38	1931	0:06	RENDEZVOUS
14 DEC	190.01	1954	0.05	DOCKING
16 DEC	236:53	1846	0:32	POST TEI
17 DEC	257:26	1519	1:04	CMP EVA
18 DEC	284:07	1800	0:30	PRESS CONFERENCE

*LSPE CHARGES AND LM ASCENT STAGE IMPACT TELEVISION TIMES ARE NOT SHOWN

PHOTOGRAPHIC EQUIPMENT

Still and motion pictures will be made of most space-craft maneuvers and crew lunar surface activities. During lunar surface operations, emphasis will be on documenting placement of lunar surface experiments, documenting lunar samples, and on recording in their natural state the lunar surface features.

Command module lunar orbit photographic tasks and experiments include high-resolution photography to aid exploration, photography of surface features of special scientific interest and astronomical phenomena such as solar corona, zodiacal light, and galactic poles.

Camera equipment stowed in the Apollo 17 command module consists of one 70mm Hasselblad electric camera, a 16mm Maurer motion picture camera, and a 35mm Nikon F single-lens reflex camera. The command module Hasselblad electric camera is normally fitted with an 80mm f/2.8 Zeiss Planar lens, but a bayonet-mount 250mm Zeiss Sonnar lens can be fitted for long-distance Earth/Moon photos.

The 35mm Nikon F is fitted with a 55mm f/1.2 Nikkor lens for the dim-light photographic experiments.

The Maurer 16mm motion picture camera in the command module has Kern-Switar lenses of 10,18 and 75mm focal length available. Accessories include a right-angle mirror, a power cable and a sextant adapter which allows the camera to film through the navigation sextant optical system.

Cameras stowed in the lunar module are two 70mm Hassel-blad data cameras fitted with 60mm Zeiss Metric lenses, an electric Hasselblad camera with 500mm lens, and one 16mm Maurer motion picture camera with 10mm lenses.

The LM Hasselblads have crew chest mounts that fit dove-tail brackets on the crewman's remote control unit, thereby leaving both hands free. The LM motion picture cameras will be mounted in the right-hand window to record descent, land-ing, ascent and rendezvous.

Descriptions of the 24-inch panoramic camera and the 3-inch mapping/stellar camera are in the orbital science section of this press kit.

ASTRONAUT EQUIPMENT

Space Suit

Apollo crewmen wear two versions of the Apollo space suit: the command module pilot version (A-7LB-CMP) for operations in the command module and for extravehicular operations during SIM bay film retrieval during transearth coast; and the extravehicular version (A-7LB-EV) worn by the commander and lunar module pilot for lunar surface EVAs.

The A-7LB-EV suit differs from Apollo suits flown prior to Apollo 15 by having a waist joint that allows greater mobility while the suit is pressurized -- stooping down for setting up lunar surface experiments, gathering samples and for sitting on the lunar roving vehicle.

From the inside out, an integrated thermal meteroid suit cover layer worn by the commander and lunar module pilot starts with rubber-coated nylon and progresses outward with layers of nonwoven Dacron, aluminized Mylar film and Beta marquisette for thermal radiation protection and thermal spacers, and finally with a layer of nonflammable Teflon-coated Beta cloth and an abrasion-resistant layer of Teflon fabric -- a total of 18 layers.

Both types of the A-7LB suit have a pressure retention portion called a torso limb suit assembly consisting of neoprene coated nylon and an outer structural restraint layer.

The space suit with gloves, and dipped rubber convolutes which serve as the pressure layer, liquid cooling garment, portable life support system (PLSS), oxygen purge system, lunar extravehicular visor assembly (LEVA), and lunar boots make up the extravehicular mobility unit (EMU). The EMU provides an extravehicular crewman with life support for a 7-hour period outside the lunar module without replenishing expendables.

Lunar extravehicular visor assembly - The assembly consists of polycarbonate shell and two visors with thermal control and optical coatings on them. The EVA visor is attached over the pressure helmet to provide impact, micro-meteoroid, thermal and ultraviolet-infrared light protection to the EVA crewmen.

After Apollo 12, a sunshade was added to the outer portion of the LEVA in the middle portion of the helmet rim.

Extravehicular gloves - Built of an outer shell of Chromel-R fabric and thermal insulation the gloves provide protection when handling extremely hot and cold objects. The finger tips are made of silicone rubber to provide more sensitivity.

Constant-wear garment - A one-piece constant-wear garment, similar to "long johns", is worn as an undergarment for the space suit in intravehicular and on CSM extravehicular operations, and with the inflight coveralls. The garment is porous-knit cotton with a waist-to-neck zipper for donning. Biomedical harness attach points are provided.

Liquid-cooling garment - The knitted nylon-spandex garment includes a network of plastic tubing through which cooling water from the PLSS is circulated. It is worn next to the skin and replaces the constant-wear garment during lunar surface EVA.

Portable life support system (PLSS) - The backpack supplies oxygen at 3.7 psi and cooling water to the liquid cooling garment. Return oxygen is cleansed of solid and gas contaminants by a lithium hydroxide and activated charcoal canister. The PLSS includes communications and telemetry equipment, displays and controls, and a power supply. The PLSS is covered by a thermal insulation jacket, (two stowed in LM).

Oxygen purge system (OPS) - Mounted atop the PLSS, the oxygen purge system provides a contingency 30-75 minute supply of gaseous oxygen in two bottles pressurized to 5,880 psia, (a minimum of 30 minutes in the maximum flow rate and 75 minutes in the low flow rate). The system may also be worn separately on the front of the pressure garment assembly torso for contingency EVA transfer from the LM to the CSM or behind the neck for CSM EVA. It serves as a mount for the VHF antenna for the PLSS, (two stowed in LM).

Coveralls - During periods out of the space suits, crewmen wear two-piece Teflon fabric inflight coveralls for warmth and for pocket stowage of personal items.

Communications carriers - "Snoopy hats" with redundant microphones and earphones are worn with the pressure helmet; a light weight headset is worn with the inflight coveralls.

-more-

Water Bags - .9 liter (1 quart) drinking water bags are attached to the inside neck rings of the EVA suits. The crewman can take a sip of water from the 6x8-inch bag through a 1/8-inch-diameter tube within reach of his mouth. The bags are filled from the lunar module potable water dispenser.

Buddy Secondary Life Support System - A connecting hose system which permits a crewman with a failed PLSS to share cooling water in the other crewman's PLSS. The BSLSS lightens the load on the oxygen purge system in the event of a total PLSS failure in that the OPS would supply breathing and pressurizing oxygen while the metabolic heat would be removed by the shared cooling water from the good PLSS. The BSLSS will be stowed on the LRV.

Lunar Boots - The lunar boot is a thermal and abrasion protection device worn over the inner garment and boot assemblies. It is made up of layers of several different materials beginning with Teflon coated beta cloth for the boot liner to Chromel R metal fabric for the outer shell assembly. Aluminized Mylar, Nomex felt, Dacron, Beta cloth and Beta marquisette Kapton comprise the other layers. The lunar boot sole is made of high-strength silicone rubber.

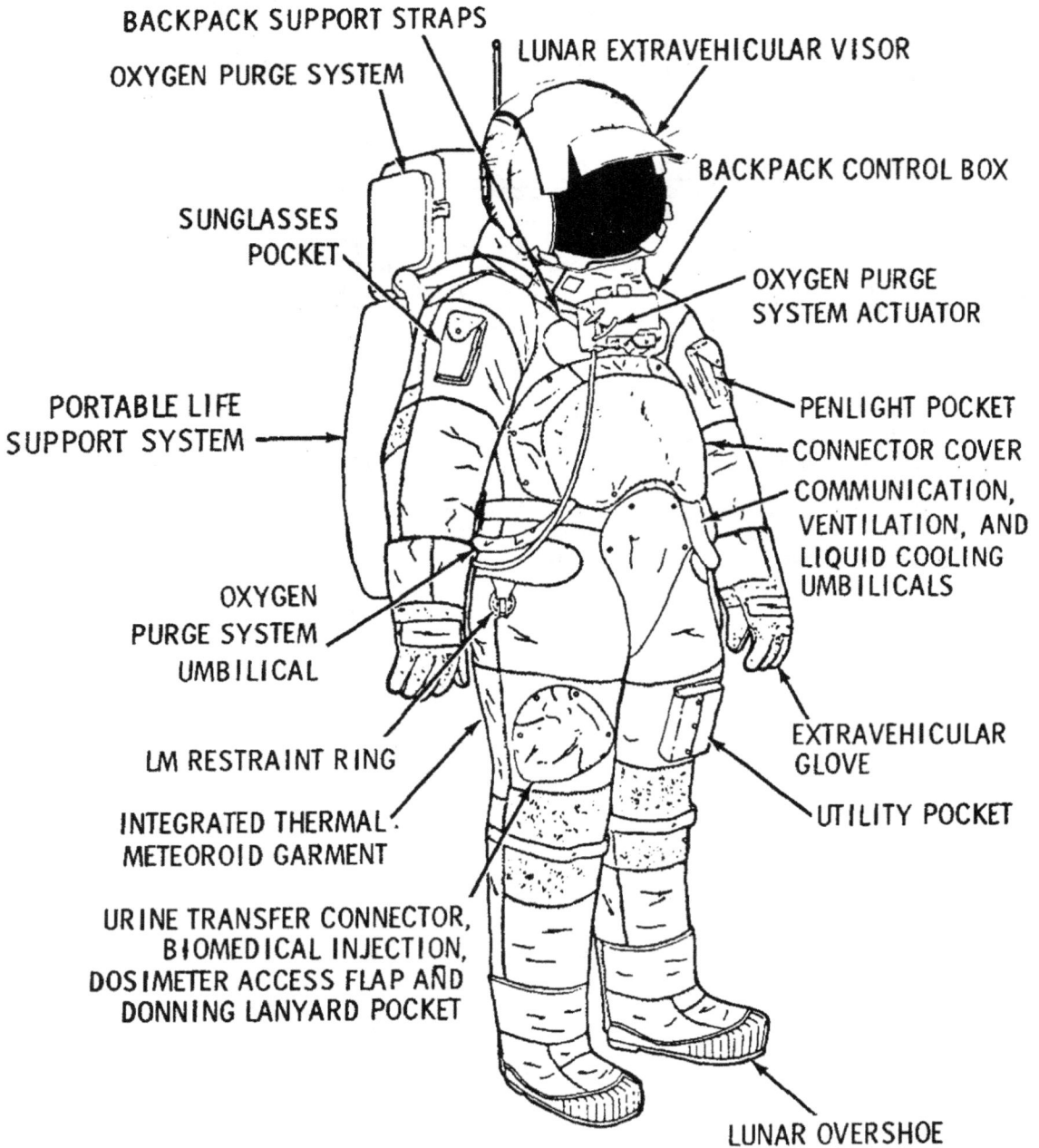

BACKPACK SUPPORT STRAPS

OXYGEN PURGE SYSTEM

LUNAR EXTRAVEHICULAR VISOR

BACKPACK CONTROL BOX

SUNGLASSES POCKET

OXYGEN PURGE SYSTEM ACTUATOR

PORTABLE LIFE SUPPORT SYSTEM

PENLIGHT POCKET

CONNECTOR COVER

COMMUNICATION, VENTILATION, AND LIQUID COOLING UMBILICALS

OXYGEN PURGE SYSTEM UMBILICAL

EXTRAVEHICULAR GLOVE

UTILITY POCKET

LM RESTRAINT RING

INTEGRATED THERMAL METEOROID GARMENT

URINE TRANSFER CONNECTOR, BIOMEDICAL INJECTION, DOSIMETER ACCESS FLAP AND DONNING LANYARD POCKET

LUNAR OVERSHOE

-more-

EXTRAVEHICULAR MOBILITY UNIT

Personal Hygiene

Crew personal hygiene equipment aboard Apollo 17 includes body cleanliness items, the waste management system, and one medical kit.

Packaged with the food are a toothbrush and a two-ounce tube of toothpaste for each crewman. Each man-meal package contains a 3.5-by-4-inch wet-wipe cleansing towel. Additionally, three packages of 12-by-12-inch dry towels are stowed beneath the command module pilot's couch. Each package contains seven towels. Also stowed under the command module pilot's couch are seven tissue dispensers containing 53 three-ply tissues each.

Solid body wastes are collected in plastic defecation bags which contain a germicide to prevent bacteria and gas formation. The bags are sealed after use, identified, and stowed for return to Earth for post-flight analysis.

Urine collection devices are provided for use while wearing either the pressure suit or the inflight coveralls. The urine is dumped overboard through the spacecraft urine dump valve in the CM and stored in the LM. On Apollo 16 urine specimens will be returned to Earth for analysis.

Survival Kit

The survival kit is stowed in two rucksacks in the right-hand forward equipment bay of the CM above the lunar module pilot.

Contents of rucksack No. 1 are: two combination survival lights, one desalter kit, three pairs of sunglasses, one radio beacon, one spare radio beacon battery and spacecraft connector cable, one knife in sheath, three water containers, two containers of Sun lotion, two utility knives, three survival blankets and one utility netting.

Rucksack No. 2: one three-man life raft with CO_2 inflater, one sea anchor, two sea dye markers, three sunbonnets, one mooring lanyard, three manlines and two attach brackets.

The survival kit is designed to provide a 48-hour post-landing (water or land) survival capability for three crewmen between 40 degrees North and South latitudes.

Medical Kits

The command module crew medical supplies are contained in two kits. Included in the larger medical accessories kit are antibiotic ointment, skin cream, eye drops, nose drops, spare biomedical harnesses, oral thermometer and pills of the following types: 18 pain, 12 stimulant, 12 motion sickness, 48 diarrhea, 60 decongestant, 21 sleeping, 72 aspirin and 60 each of two types of antibiotic. A smaller command module auxiliary drug kit contains 80 and 12 of two types of pills for treatment of cardiac arrythymia and two injectors for the same symptom.

The lunar module medical kit contains eye drops, nose drops, antibiotic ointment, bandages and the following pills: 4 stimulant, 4 pain, 8 decongestant, 12 diarrhea, 12 aspirin and 6 sleeping. A smaller kit in the LM contains 8 and 4 units of injectable drugs for cardiac arrythymia and 2 units for pain suppression.

Crew Food System

The Apollo 17 crew selected menus for their flight from the largest variety of foods ever available for a U.S. manned mission. As on Apollo 16, the preflight, inflight, and postflight diets are being monitored to facilitate interpretation of the medical tests.

Menus were designed upon individual crewmember physiological requirements in the unique conditions of weightlessness and one-sixth gravity on the lunar surface. Daily menus provide approximately 2500 calories per day for each crewmember.

Food items are assembled into meal units and identified as to crewmember and sequence of consumption. Foods stored in the "pantry" may be used as substitutions for nominal meal items so long as the nutrient intake for a 24-hour period is not altered significantly.

There are various types of food used in the menus. These include freeze-dried rehydratables in spoon-bowl packages; thermostabilized foods (wet packs) in flexible packages and metal easy-open cans; intermediate moisture foods; dry bite-size cubes; and beverages.

Water for drinking and rehydrating food is obtained from two sources in the Command Module -- a portable dispenser for drinking water and a water spigot at the food preparation station which supplies water at about 145 degrees and 55 degrees Fahrenheit. The potable water dispenser provides a continuous flow of water as long as the trigger is held down, while the food preparation spigot dispenses water in one-ounce increments.

A continuous-flow water dispenser similar to the one in the Command Module is used aboard the Lunar Module for cold water reconstitution of food stowed aboard the Lunar Module.

Water is injected into a food package and the package is kneaded and allowed to sit for several minutes. The bag top is then cut to open and the food eaten with a spoon. After a meal, germicide tablets are placed in each bag to prevent fermentation of any residual food and gas formation. The bags are then rolled and stowed in waste disposal areas in the spacecraft.

-more-

An improved Skylab beverage package design will be used by the crew to measure water consumption. Functional aspects of the package and the behavior of liquid during extended periods of weightlessness will be observed.

The in-suit drink device will contain water as on the Apollo 15 mission. As on Apollo 15 and 16, the crewmen on the lunar surface will have the option to snack on an in-suit food bar.

The nutritionally complete fruitcake provides all the nutrients needed by man in their correct proportions. The fruitcake contains many ingredients such as: soy flour, wheat flour, sugar, eggs, salt, cherries, pineapple, nuts, raisins, and shortening. Vitamins have been added. The product is heat sterilized in an impermeable flexible pouch and is shelf-stable until opened. This fruitcake can provide a nutritious snack or meal. This food is planned for use in the future in the Space Shuttle program as a contingency food system.

The irradiated ham provides the crew with a shelf-stable slice of ham 12 mm thick. Each slice weighs about 100 grams and may be used for making sandwiches during flight. The radiation sterilization (radappertization) is performed while the ham is at -40°C. The absorbed irradiation dose is 3.7 to 4.3 million rads. This gives an excellent product with an expected shelf-life of 3 years.

The fruitcake and ham slices were specially developed and provided for Apollo 17 by the U.S. Army Natick Laboratories, Natick, Massachusetts.

New foods for the Apollo 17 mission are irradiated sterilized ham, nutrient complete fruitcake, and rehydratable tea and lemonade beverages.

APOLLO 17 CSM MENU Eugene A. Cernan, CDR (Red Velcro)

MEAL A

Day 1*, 5, 9***, 13		Day 2, 6**, 10, 14**		Day 3, 11		Day 4, 12	
Bacon Squares (8)	IMB	Spiced Oat Cereal	RSB	Scrambled Eggs	RSB	Sausage Patties	R
Scrambled Eggs	RSB	Sausage Patties	R	Bacon Squares (8)	IMB	Apricot Cereal Cubes (4)	DB
Cornflakes	RSB	Mixed Fruit	WP	Peaches	WP	Fruit Cocktail	R
Peaches	RSB	Cinnamon Toast Bread (4)	DB	Pineapple GF Drink	R	Pears	IMB
Orange Beverage	R	Instant Breakfast	R	Cocoa w/K	R	Cocoa w/K	R
Cocoa	R	Coffee w/K	R			Coffee	R

MEAL B

Day 1*, 5, 9***, 13		Day 2, 6**, 10, 14**		Day 3, 11		Day 4, 12	
Chicken & Rice Soup	RSB	Corn Chowder	RSB	Lobster Bisque	RSB	Chicken Soup	RSB
Meatballs and Sauce	WP	Frankfurters	WP	Peanut Butter	WP	Ham (Ir)	WP
Fruitcake	WP	Bread, white (2)	WP	Jelly	WP	Cheddar Cheese Spread	WP
Lemon Pudding	WP	Catsup	WP	Bread, white (1)	WP	Bread, Rye (1)	
Orange P/A Drink	R	Apricots	IMB	Chocolate Bar	IMB	Cereal Bar	IMB
		Orange GF Drink	R	Orange GF Drink w/K	R	Orange Beverage	R

MEAL C

Day 1*, 5, 9***, 13		Day 2, 6**, 10, 14**		Day 3, 11		Day 4, 12	
Potato Soup	RSB	Turkey and Gravy	WP	Shrimp Cocktail	WP	Tomato Soup	RSB
Beef and Gravy	WP	Pork & Potatoes	RSB	Beef Steak	RSB	Hamburger	WP
Chicken Stew	RSB	Brownies (4)	DB	Butterscotch Pudding	DB	Mustard	WP
Ambrosia, Peach	RSB	Orange Juice	R	Peaches	R	Vanilla Pudding	WP
Gingerbread (4)	DB	Lemonade	R	Orange Drink w/K	R	Date Fruitcake (4)	IMB
Citrus Beverage	R					Orange P/A Drink w/K	R

*   Meal C only
**   Meal A only
*** Meals B and C only

DB = Dry Bite   IMB = Intermediate Moisture Bite
R = Rehydratable   RSB = Rehydratable Spoon Bowl
WP = Wet Pack   Ir = Irradiated

APOLLO 17 — LM MENU, Eugene A. Cernan, CDR (Red Velcro)

MEAL B

Day 6		Day 7		Day 8		Day 9	
A		A		A		A	
Corn Chowder	RSB	Scrambled Eggs	RSB	Sausage Patties	R	Bacon Squares (8)	IMB
Franfurters (2)	WP	Bacon Squares (8)	IMB	Apricot Cereal Cubes (6)	DB	Scrambled Eggs	RSB
Bread, white	WP	Peaches	IMB	Fruit Cocktail	R	Cornflakes	RSB
Catsup	IMB	Peanut Butter	WP	Pears	IMB	Beef and Gravy	WP
Apricots	R	Jelly	WP	Cereal Bar	IMB	Fruitcake	WP
Orange GF Drink	R	Bread, white (1)	IMB	Cheese Cracker Cube (4)	DB	Peaches	RSB
Tea		Chocolate Bar	R	Ham (Ir)	WP	Cocoa	R
Lemonade		Pineapple GF Drink	R	Cocoa	R	Orange Beverage	R
		Orange GF Drink w/K	R	Tea	R	Tea	R
		Cocoa	R	Spiced Oat Cereal	RSB		
		Tea	R	Lemonade	R		

MEAL C

Day 6		Day 7		Day 8		Day 9	
B		B		B			
Spaghetti & Meat Sauce	RSB	Chicken and Rice	RSB	Lobster Bisque	RSB		
Turkey and Gravy	WP	Shrimp Cocktail	WP	Hamburger	WP		
Pork and Potatoes	RSB	Beef Steak	RSB	Mustard	WP		
Brownies (4)	DB	Beef Sanwiches (4)	DB	Cheddar Cheese Spread	WP		
Orange Beverage	R	Butterscotch Pudding	RSB	Bread, rye (1)	IMB		
Tea	R	Graham Cracker Cube (6)	DB	Date Fruitcake (4)	IMB		
		Orange Drink w/K	R	Orange PA Drink w/K	R		
		Tea	R	Orange Beverage	R		
				Tea	R		

In-Suit Food Bar Assembly	6 ea	P/N:	SEB 13100318-301
In-Suit Drinking Device	4 ea	P/N:	14-0151-02
Spoon Assembly (2)	1 ea	P/N:	14-0144-01
Germicidal Tablets Pouch (42)	1 ea	P/N:	14-02166
Germicidal Tablets Pouch (20)	1 ea	P/N:	14-

DB = Dry Bite   IMB = Intermediate Moisture Bite
R = Rehydratable   RSB = Rehydratable Spoon Bowl
WP = Wet Pack

-more-

APOLLO 17 CSM MENU, Harrison H. Schmitt, LMP (Blue Velcro)

MEAL	Day 1*, 5, 9***, 13	Day 2, 6***, 10, 14**	Day 3, 11	Day 4, 12
A	Bacon Squares (8) IMB Scrambled Eggs RSB Cornflakes RSB Apricots IMB Cocoa R	Sausage Patties R Cinnamon Toast Bread (4) DB Mixed Fruit WP Instant Breakfast R Coffee w/K R	Scrambled Eggs RSB Bacon Squares (8) IMB Peaches WP Orange P/A Drink w/K R Cocoa R	Sausage Patties R Grits RSB Peaches RSB Pears IMB Pineapple GF Drink R Coffee w/K R
B	Chicken & Rice Soup RSB Meatballs w/Sauce WP Fruitcake WP Lemon Pudding WP Citrus Beverage R	Corn Chowder RSB Frankfurters WP Bread, White (2) WP Catsup RSB Chocolate Pudding RSB Orange GF Drink w/K R	Potato Soup RSB Peanut Butter WP Jelly WP Bread, White (1) IMB Cherry Bar (1) IMB Orange GF Drink w/K R	Chicken Soup RSB Ham (Ir) WP Cheddar Cheese Spread WP Bread, Rye (1) IMB Cereal Bar IMB Orange Drink w/K R
C	Lemonade R Beef & Gravy WP Chicken Stew RSB Ambrosia RSB Gingerbread (4) DB Grapefruit Drink R	Turkey & Gravy WP Pork and Potatoes RSB Carmel Candy IMB Orange Juice R	Shrimp Cocktail RSB Beef Steak WP Butterscotch Pudding RSB Peaches IMB Orange Drink w/K R	Tomato Soup RSB Hamburger WP RSB Mustard WP Vanilla Pudding WP Chocolate Bar IMB Grape Drink w/K R

CALORIES

* Meal C only
** Meal A only
*** Meal B and C only

DB = Dry Bite
R = Rehydratable
WP = Wet Pack

IMB = Intermediate Moisture Bite
RSB = Rehydratable Spoon Bowl
Ir = Irradiated

LM Menu Continued

APOLLO 17 - LM MENU, Harrison H. Schmitt, LMP (Blue Velcro)

MEAL	Day 6		Day 7		Day 8		Day 9	
B	Corn Chowder	RSB	A Scrambled Eggs	RSB	A Sausage Patties	R	A Bacon Squares (8)	IMB
	Frankfurters	WP	Bacon Squares (8)	IMB	Spiced Oat Cereal	RSB	Scrambled Eggs	RSB
	Bread, White (2)		Peaches	IMB	Peaches	RSB	Cornflakes	RSB
	Catsup	WP	Peanut Butter	WP	Pears	IMB	Apricots	IMB
	Chocolate Pudding	RSB	Jelly	WP	Cereal Bar	IMB	Cocoa	R
	Orange GF Drink	R	Bread, White (1)		Gingerbread (6)	DB	Tea	R
	Tea	R	Orange GF Drink w/K	R	Ham (Ir)	WP	Beef and Gravy	WP
	Lemonade	R	Cocoa	R	Pineapple GF Drink	R	Fruitcake	WP
			Tea	R	Tea	R		
			Fruit Cocktail	R				
C	Turkey and Gravy	WP	B Chicken & Rice	RSB	B Potato Soup	RSB		
	Pork and Potatoes	RSB	Shrimp Cocktail	RSB	Hamburger	WP		
	Carmel Candy	IMB	Beef Steak	WP	Mustard	WP		
	Orange Beverage	R	Beef Sandwiches (4)	DB	Cheddar Cheese Spread	WP		
	Tea	R	Butterscotch Pudding	RSB	Bread, Rye (1)			
			Graham Cracker Cube (6)	DB	Chocolate Bar	IMB		
			Orange Drink w/K	R	Banana Pudding	RSB		
			Orange P/A Drink	R	Orange Drink w/K	R		
			Tea	R	Grape Drink w/K	R		
					Tea	R		

more

APOLLO 17 CSM MENU, RONALD E. EVANS, CMP (White Velcro)

MEAL A

Day 1*,5,9,13		Day 2,6,10,14**		Day 3,7,11		Day 4,8,12	
Bacon Squares (8)	IMB	Spiced Oat Cereal	RSB	Scrambled Eggs	RSB	Sausage	R
Scrambled Eggs	RSB	Sausage Patties	R	Bacon Squares (8)	IMB	Grits	RSB
Cornflakes	RSB	Mixed Fruit	WP	Peaches	WP	Fruit Cocktail	R
Apricots	IMB	Instant Breakfast	R	Cinnamon Toast Bread(4)	DB	Orange Beverage	R
Orange Juice	R	Coffee w/K	R	Orange Juice	R	Coffee w/K	R
				Cocoa w/K	R		

MEAL B

Day 1*,5,9,13		Day 2,6,10,14		Day 3,7,11		Day 4,8,12	
Chicken & Rice Soup	RSB	Franfurters	WP	Lobster Bisque	RSB	Ham (Ir)	WP
Meatballs w/Sauce	WP	Bread, white (2)	WP	Peanut Butter	WP	Cheddar Cheese Spread	WP
Fruitcake	WP	Catsup	WP	Jelly	WP	Bread, rye (1)	
Butterscotch Pudding	WP	Pears	IMB	Bread, white (1)		Peaches	RSB
Orange PA Drink	R	Chocolate Pudding	RSB	Cherry Bar (1)	IMB	Cereal Bar	IMB
		Grape Drink w/K	R	Citrus Beverage w/K	R	Orange PA Drink w/K	R

MEAL C

Day 1*,5,9,13		Day 2,6,10,14		Day 3,7,11		Day 4,8,12	
Potato Soup	RSB	Corn Chowder	RSB	Shrimp Cocktail	RSB	Tomato Soup	RSB
Beef and Gravy	WP	Turkey & Gravy	WP	Beef Steak	WP	Hamburger	WP
Chicken Stew	RSB	Chocolate Bar	IMB	Butterscotch Pudding	RSB	Mustard	WP
Ambrosia	DB	Orange Beverage	R	Orange Drink w/K	R	Vanilla Pudding	WP
Brownies (4)	R					Sugar Cookies (4)	DB
Orange GF Drink						Carmel Candy	IMB
						Grape Drink w/K	R

*   Meal C only
** Meal A only

DB = Dry Bite
R * = Rehydratable
WP = Wet Pack

IMB = Intermediate Moisture Bite
RSB = Rehydratable Spoon Bowl
Ir = Irradiated

-more-

APOLLO 17

PANTRY STOWAGE ITEMS

BEVERAGES	QTY.	ACCESSORIES	QTY.
Coffee (B)	20	Contingency Feeding System	1
Tea	20		
Grape Drink	10	Germicidal Tablets (42)	3
Grape Punch	10		
		Index Card	1
		S/L Beverage Dispenser (empty)	3
		Contingency Beverages	30
		(For Contingency Use Only)	

    15 Instant Breakfast
     5 Orange Drink
     5 Pineapple Orange Drink
     5 Lemonade

SNACK ITEMS

Bacon Squares (4)	9
Apricot Cereal Cubes (4)	6
Brownies (4)	3
Gingerbread (4)	3
Graham Crackers (4)	6
Jellied Candy	6
Peach Ambrosia	3
Pecans (6)	6
Fruitcake (WP)	3
Sugar Cookies (4)	6
Apricots (IMB)	3
Peaches (IMB)	3
Pears (IMB)	3
Chocolate Bar (IMB)	3
Tuna Salad Spread (WP) (Small Cans)	2
Catsup (WP)	3

SATURN V LAUNCH VEHICLE

The Saturn V launch vehicle (SA-512) assigned to the
Apollo 17 mission is similar to the vehicles used for the
missions of Apollo 8 through Apollo 16.

First Stage

The five first stages (S-1C) F-1 engines develop about
34 million newtons (7.67 million pounds) of thrust at launch.
Major stage components are the forward skirt, oxidizer tank,
intertank structure, fuel tank, and thrust structure.
Propellant to the five engines normally flows at a rate of
about 13,200 kilograms (29,200 pounds; 3,370 gallons) a second.
One engine is rigidly mounted on the stage's centerline; the
outer four engines are mounted on a ring equally spaced around
the center engine. These outer engines are gimbaled to control
the vehicle's attitude during flight.

Second Stage

The five second stage (S-II) J-2 engines develop a total
of about 5.13 million newtons (1.15 million pounds) of thrust
during flight. Major components are the forward skirt, liquid
hydrogen and liquid oxygen tanks (separated by an insulated
common bulkhead) a thrust structure, and a interstage section
that connects the first and second stages. The engines are
mounted and used in the same arrangement as the first stage's
F-1 engines: four outer engines can be gimbaled; the center
one is fixed.

Third Stage

Major components of the third stage (S-IVB) are a single
J-2 engine, aft interstage and skirt, thrust structure, two
propellant tanks with a common bulkhead, and forward skirt.
The gimbaled engine has a maximum thrust of .93 million
newtons (209,000 pounds), and can be restarted in Earth orbit.

Instrument Unit

The instrument unit (IU) contains navigation, guidance and control equipment to steer the Saturn V into Earth orbit and translunar trajectory. The six major systems are structural, enviromental control, guidance and control, measuring and telemetry, communications, and electrical.

The IU's inertial guidance platform provides space-fixed reference coordinates and measures acceleration during flight. If the platform should fail during boost, systems in the Apollo spacecraft are programmed to provide launch vehicle guidance. After second stage ignition, the spacecraft commander can manually steer the vehicle if its guidance platform is lost.

Propulsion

The Saturn V has 31 propulsive units, with thrust ratings ranging from 311 newtons (70 pounds) to more than 6.8 million newtons (1.53 million pounds). The large main engines burn liquid propellants; the smaller units use solid or hypergolic (self-igniting) propellants.

The five F-1 engines give the first stage a thrust range of from 34,096,110 newtons, (7,665,111 pounds) at liftoff to 40,207,430 newtons (9,038,989 pounds) at center engine cutoff. Each F-1 engine weighs almost nine metric tons (10 short tons), is more than 5.5 meters long (18 feet), and has a nozzle exit diameter of nearly 4.6 meters (14 feet). Each engine uses almost 2.7 metric tons (3 short tons) of propellant a second.

The five J-2 engines on the second stage develop an average thrust of 5,131,968 newtons (1,153,712 pounds) during flight. The one J-2 engine of the third stage develops an average thrust of 926,307 newtons (208,242 pounds). The 1,590-kilogram (3,500-pound) J-2 engine uses high-energy, low-molecular-weight liquid hydrogen as fuel, and liquid oxygen as oxidizer.

The first stage has eight solid-fuel retro-rockets that fire to separate the first and second stages. Each rocket produces a thrust of 337,000 newtons (75,800 pounds) for 0.54 seconds.

Four retrorockets, located in the third stage's aft interstage, separate the second and third stages. Two jettisonable ullage rockets settle propellants before engine ignition. Six smaller engines in two auxiliary propulsion system modules on the third stage provide three-axis attitude control.

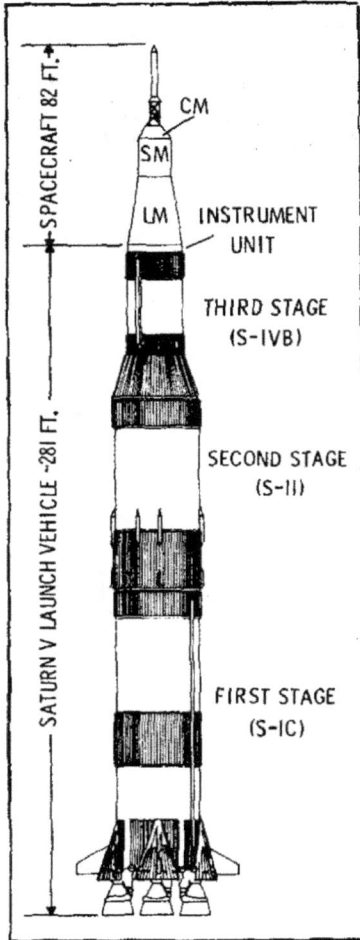

INSTRUMENT UNIT (IU)

Diameter: 6.6 meters (21.7 feet)
Height: 0.9 meters (3 feet)
Weight: 2,040 kilograms (4,500 pounds)

THIRD STAGE (S-IVB)

Diameter: 6.6 meters (21.7 feet)
Height: 18.1 meters (59.3 feet)
Weight: 121,000 kg. fueled (266,000 lbs.)
11,300 kg.dry (24,900 lbs.)
Engine: One J-2
Propellants: Liquid Oxygen 89,000 kg. (196,000 lbs.)
Liquid Hydrogen 19,900 kg. (43,750 lbs.)
Thrust: 926,367 newtons (208,242 lbs.)
Interstage: 3,637 kg. (8,019 lbs.)

SECOND STAGE (S-II)

Diameter: 10.1 meters (33 feet)
Height: 24.8 meters (81.5 feet)
Weight: 493,318 kg.fueled (1,087,580 lbs.)
36,478 kg.dry (80,420 lbs.)
Engines: Five J-2
Propellants: Liquid Oxygen 384,000 kg. (845,713 lbs.)
Liquid Hydrogen 73,000 kg. (160,464 lbs.)
Thrust: 5,131,968 newtons (1,153,712 lbs.)
Interstage: 4,541 kg. (9,990 lbs.)

FIRST STAGE (S-IC)

Diameter: 10.1 meters (33 feet)
Height: 42.1 meters (138 feet)
Weight: 2,246.540 kg.fueled (4,952,775 lbs.)
130,441 kg dry (287,574 lbs.)
Engines: Five F-1
Propellants: Liquid Oxygen 1,471,427 kg.
(3,243,942 lbs.) RP-1 Kerosene
642,177 kg. (1,415,257 lbs.)
Thrust: 34,096,110 newtons (7,665,111 lbs.)
at lift-off

NOTE: Weights and measures given above are for the nominal vehicle configuration for Apollo 17. The figures may vary slightly due to changes before launch to meet changing conditions. Weights of dry stages and propellants do not equal total weight because frost and miscellaneous smaller items are not included in chart.

SATURN V LAUNCH VEHICLE

-more-

APOLLO SPACECRAFT

The Apollo spacecraft consists of the command module, service module, lunar module, a spacecraft lunar module adapter (SLA), and a launch escape system. The SLA houses the lunar module and serves as a mating structure between the Saturn V instrument unit and the SM.

Launch Escape System (LES) -- The function of the LES is to propel the command module to safety in an aborted launch. It has three solid-propellant rocket motors: a 658,000 newton (147,000-pound)-thrust launch escape system motor, a 10,750-newton (2,400-pound)-thrust pitch control motor, and a 141,000 newton (31,500-pound)-thrust tower jettison motor. Two canard vanes deploy to turn the command module aerodynamically to an attitude with the heat-shield forward. The system is 10 meters (33 feet) tall and 1.2 meters (four feet) in diameter at the base, and weighs 4,158 kilograms (9,167 pounds).

Command Module (CM) -- The command module is a pres-sure vessel encased in heat shields, cone-shaped, weighing 5,843.9 kg (12,874 lb.) at launch.

The command module consists of a forward compartment which contains two reaction control engines and components of the Earth landing system; the crew compartment or inner pressure vessel containing crew accommodations, controls and displays, and many of the spacecraft systems; and the aft compartment housing ten reaction control engines, propellant tankage, helium tanks, water tanks, and the CSM umbilical cable. The crew compartment contains 6 cubic meters (210 cubic ft.) of habitable volume.

Heat-shields around the three compartments are made of brazed stainless steel honeycomb with an outer layer of phenolic epoxy resin as an ablative material.

-    The CSM and LM are equipped with the probe-and-drogue docking hardware. The probe assembly is a powered folding coupling and impact attentuating device mounted in the CM tunnel that mates with a conical drogue mounted in the LM docking tunnel. After the 12 automatic docking latches are checked following a docking maneuver, both the probe and drogue are removed to allow crew transfer between the CSM and LM.

- more -

COMMAND MODULE

SERVICE MODULE

Service Module (SM) -- The Apollo 17 service module will weigh 24,514 kg (54,044 lb.) at launch, of which 18,415 kg (40,594 lb.) is propellant for the 91,840 newton (20,500 pound) - thrust service propulsion engine: (fuel: 50/50 hydrazine and unsymmetrical dimethyl-hydrazine; oxidizer: nitrogen textroxide). Aluminum honeycomb panels 2.54 centimeters (one inch) thick form the outer skin, and milled aluminum radial beams separate the interior into six sections around a central cylinder containing service propulsion system (SPS) helium pressurant tanks. The six sectors of the service module house the following components: Sector I -- oxygen tank 3 and hydrogen tank 3, J-mission Scientific Instrumentation Module (SIM) bay; Sector II -- space radiator, +Y RCS package, SPS oxidizer storage tank; Sector III -- space radiator, +Z RCS package, SPS oxidizer storage tank; Sector IV -- three fuel cells, two oxygen tanks, two hydrogen tanks, auxiliary battery; Sector V -- space radiator, SPS fuel sump tank, -Y RCS package; Sector VI -- space radiator, SPS fuel storage tank, -Z RCS package.

Spacecraft-LM adapter (SLA) Structure -- The spacecraft-LM adapter is a truncated cone 8.5 m (28 ft.) long tapering from 6.7 m (21.6 ft.) in diameter at the base to 3.9 m (12.8 ft.) at the forward end at the service module mating line. The SLA weighs 1,841 kg (4,059 lb.) and houses the LM during launch and the translunar injection manuever until CSM separation, transposition, and LM extraction. The SLA quarter panels are jettisoned at CSM separation.

Lunar Module (LM)

The lunar module is a two-stage vehicle designed for space operations near and on the Moon. The lunar module stands 7 m (22 ft. 11 in.) high and is 9.5 m (31 ft.) wide (diagonally across landing gear). The ascent and descent stages of the LM operate as a unit until staging, when the ascent stage functions as a single spacecraft for rendezvous and docking with the CM.

Ascent Stage -- Three main sections make up the ascent stage: the crew compartment, midsection, and aft equipment bay. Only the crew compartment and midsection are pressurized 337.5 grams per square centimeter (4.8 pounds per square inch gauge). The cabin volume is 6.7 cubic meters (235 cubic feet). The stage measures 3.8 m (12 ft. 4 in.) high by 4.3 m (14 ft. 1 in.) in diameter. The ascent stage has six substructural areas: crew compartment, midsection, aft equipment bay, thrust chamber assembly cluster supports, antenna supports, and thermal and micro-meteoroid shield.

The cylindrical crew compartment is 2.35 m (7 ft. 10 in.) in diameter and 1.07 m (3 ft. 6 in.) deep. Two flight stations are equipped with control and display panels, armrests, body restraints, landing aids, two front windows, an overhead docking window, and an alignment optical tele-scope in the center between the two flight stations. The habitable volume is 4.5 cubic meters (160 cubic ft.)

A tunnel ring atop the ascent stage meshes with the command module docking latch assemblies. During docking, the CM docking ring and latches are aligned by the LM drogue and the CSM probe.

The docking tunnel extends downward into the mid-section 40 cm (16 in.). The tunnel is 81 cm (32 in.) in diameter and is used for crew transfer between the CSM and LM. The upper hatch on the inboard end of the docking tunnel opens inward and cannot be opened without equalizing pressure on both hatch surfaces.

A thermal and micrometeoroid shield of multiple layers of Mylar and a single thickness of thin aluminum skin encases the entire ascent stage structure.

- more -

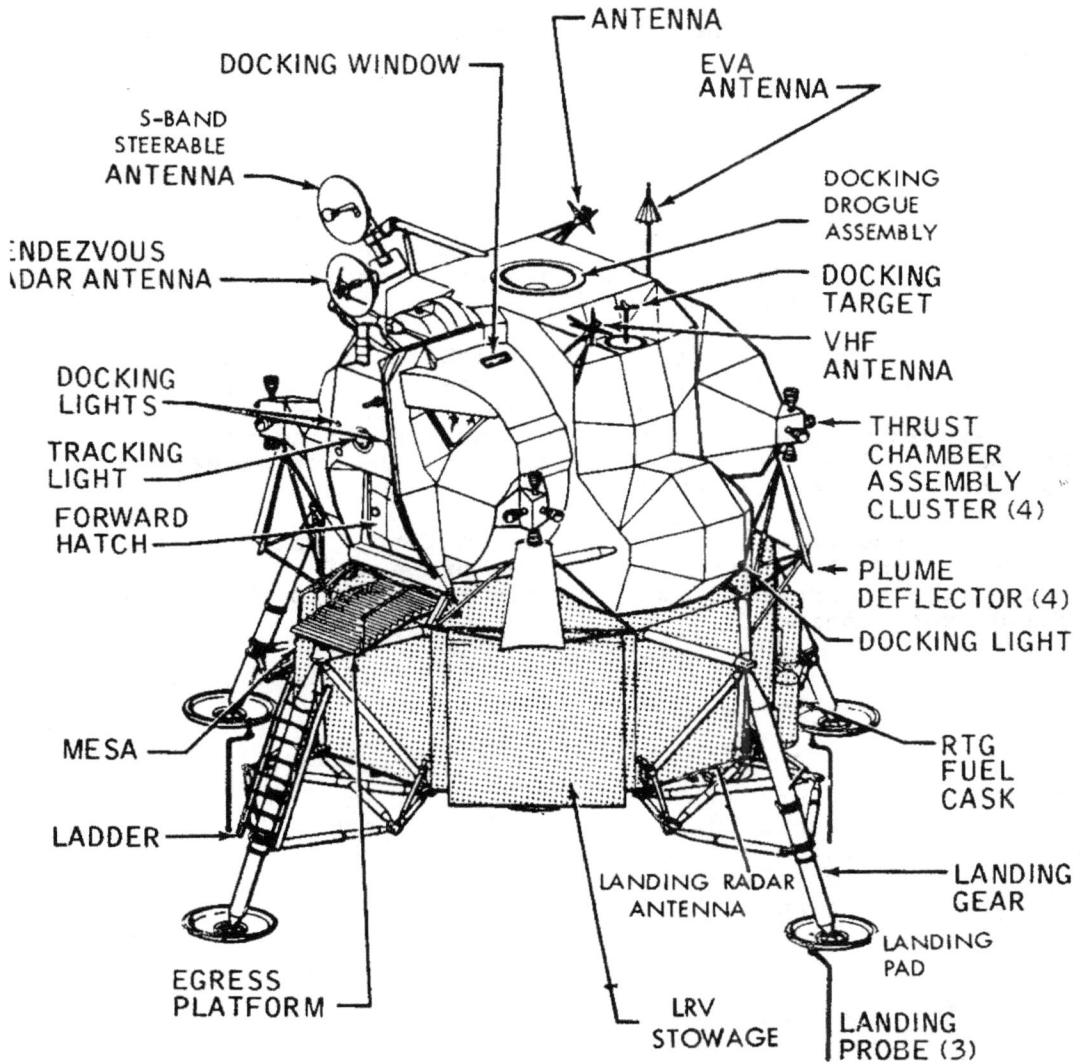

ANTENNA

DOCKING WINDOW

EVA
ANTENNA

S-BAND
STEERABLE
ANTENNA

DOCKING
DROGUE
ASSEMBLY

RENDEZVOUS
RADAR ANTENNA

DOCKING
TARGET

VHF
ANTENNA

DOCKING
LIGHTS

TRACKING
LIGHT

THRUST
CHAMBER
ASSEMBLY
CLUSTER (4)

FORWARD
HATCH

PLUME
DEFLECTOR (4)

DOCKING LIGHT

MESA

RTG
FUEL
CASK

LADDER

LANDING
GEAR

EGRESS
PLATFORM

LANDING RADAR
ANTENNA

LANDING
PAD

LRV
STOWAGE

LANDING
PROBE (3)

LUNAR MODULE

-more-

Descent Stage -- The descent stage center compartment
houses the descent engine, and descent propellant tanks
are housed in the four bays around the engine. Quadrant
II contains ALSEP. The radioisotope thermoelectric generator (RTG) is externally mounted. Quadrant IV contains the
MESA. The descent stage measures 3.2 m (10 ft. 7 in.)
high by 4.3 m (14 ft. 1 in.) in diameter and is encased
in the Mylar and aluminum alloy thermal and micrometeoroid
shield. The LRV is stowed in Quadrant I.

The LM egress platform or "porch" is mounted on the
forward outrigger just below the forward hatch. A ladder
extends down the forward landing gear strut from the porch
for crew lunar surface operations.

The landing gear struts are released explosively and
are extended by springs. They provide lunar surface landing
impact attenuation. The main struts are filled with crush-
able aluminum honeycomb for absorbing compression loads.
Footpads 0.95 m (37 in.) in diameter at the end of each
landing gear provide vehicle support on the lunar surface.

Each pad (except forward pad) is fitted with a 1.7-m
(68-in.) long lunar surface sensing probe which upon contact with the lunar surface signals the crew to shut down
the descent engine.

The Apollo LM has a launch weight of 16,429 kg (36,244)
lb.). The weight breakdown is as follows:

		kilograms	pounds
1.	Ascent stage, dry*	2,059	4,729
2.	APS propellants (loaded)	2,378	5,243
3.	Descent stage, dry	2,791	6,155
4.	DPS propellants (loaded)	8,838	19,486
5.	RCS propellants (loaded)	286	631
		16,352 kg	36,244 lbs

* Includes water and oxygen; no crew.

- more -

NATIONAL AERONAUTICS AND SPACE ADMINISTRATION

WASHINGTON, D. C. 20546

BIOGRAPHICAL DATA

NAME: Eugene A. Cernan (Captain, USN)
        NASA Astronaut   - Apollo 17 Commander

BIRTHPLACE AND DATE: Born in Chicago, Illinois, on March 14, 1934.
        His mother, Mrs. Andrew G. Cernan, resides in Bellwood,
        Illinois.

PHYSICAL DESCRIPTION: Brown hair; blue eyes; height: 6 feet;
        weight: 175 pounds.

EDUCATION: Graduated from Proviso Township High School in Maywood,
        Illinois; received a Bachelor of Science degree in Electri-
        cal Engineering from Purdue University and a Master of
        Science degree in Aeronautical Engineering from the U.S.
        Naval Postgraduate School; recipient of an Honorary Doctorate
        of Laws from Western State University College of Law in 1969
        and an Honorary Doctorate of Engineering from Purdue Univer-
        sity in 1970.

MARITAL STATUS: Married to the former Barbara J. Atchley of
        Houston, Texas.

CHILDREN: Teresa Dawn, March 4, 1963.

RECREATIONAL INTERESTS: His hobbies include horses, motorcycling,
        and all sports activities.

ORGANIZATIONS: Member of the Society of Experimental Test Pilots;
        Tau Beta Pi, national engineering society; Sigma Xi, nation-
        al science research society; and Phi Gamma Delta, national
        social fraternity.

SPECIAL HONORS: Awarded the NASA Distinguished Service Medal, the
        NASA Exceptional Service Medal, the MSC Superior Achievement
        Award, the Navy Distinguished Service Medal, the Navy Astro-
        naut Wings, the Navy Distinguished Flying Cross, the Nation-
        al Academy of Television Arts and Sciences Special Trustees
        Award (1969), and an Honorary Lifetime Membership in the
        American Federation of Radio and Television Artists.

EXPERIENCE: Cernan, a United States Navy Captain, received his
            commission through the Navy ROTC program at Purdue. He
            entered flight training upon graduation.

            He was assigned to Attack Squadrons 126 and 113 at the
            Miramar, California, Naval Air Station and subsequently
            attended the Naval Postgraduate School.

            He has logged more than 3,800 hours flying time, with more
            than 3,600 hours in jet aircraft.

CURRENT ASSIGNMENT: Captain Cernan was one of the third group
            of astronauts selected by NASA in October 1963.

            He occupied the pilot seat along side of command pilot Tom
            Stafford on the Gemini 9 mission. During this 3-day flight
            which began on June 3, 1966, the spacecraft attained a cir-
            cular orbit of 161 statute miles; the crew used three dif-
            ferent techniques to effect rendezvous with the previously
            launched Augmented Target Docking Adapter; and Cernan logged
            two hours and ten minutes outside the spacecraft in extra-
            vehicular activity. The flight ended after 72 hours and
            20 minutes with a perfect reentry and recovery as Gemini
            9 landed within 1 1/2 miles of the prime recovery ship
            USS WASP and 3/8 of a mile from the predetermined target

            Cernan subsequently served as backup pilot for Gemini 12
            and as backup lunar module pilot for Apollo VII.

            He was lunar module pilot on Apollo X, May 18-26, 1969,
            the first comprehensive lunar-orbital qualification and
            verification flight test of an Apollo lunar module. He
            was accompanied on the 248,000 nautical mile sojourn to
            the Moon by Thomas P. Stafford (spacecraft commander) and
            John W. Young (command module pilot). In accomplishing
            all of the assigned objectives of this mission, Apollo X
            confirmed the operational performance, stability, and
            reliability of the command/service module/lunar module
            configuration during translunar coast, lunar orbit inser-
            tion, and lunar module separation and descent to within 8
            nautical miles of the lunar surface. The latter maneuver
            involved employing all but the final minutes of the
            technique prescribed for use in an actual lunar landing,
            and completing critical evaluations of the lunar module
            propulsion systems and rendezvous and landing radar devices
            in subsequent rendezvous and re-docking maneuvers. In
            addition to demonstrating that man could navigate safely
            and accurately in the Moon's gravitational fields, Apollo X
            photographed and mapped tentative landing sites for future
            missions. This was Captain Cernan's second space flight
            giving him more than 264 hours and 24 minutes in space.
            Captain Cernan has since served as backup spacecraft command-
            er for the Apollo XIV flight.

NATIONAL AERONAUTICS AND SPACE ADMINISTRATION

WASHINGTON, D. C. 20546

BIOGRAPHICAL DATA

NAME: Ronald E. Evans (Commander, USN)
NASA Astronaut - Apollo 17 command module pilot

BIRTHPLACE AND DATE: Born November 10, 1933, in St. Francis, Kansas. His father, Mr. Clarence E. Evans, lives in Bird City, Kansas, and his mother, Mrs. Marie A. Evans, resides in Topeka, Kansas.

PHYSICAL DESCRIPTION: Brown hair; brown eyes; height: 5 feet 11 1/2 inches; weight: 160 pounds.

EDUCATION: Graduated for Highland Park High School in Topeka, Kansas; received a Bachelor of Science degree in Electrical Engineering from the University of Kansas in 1956 and a Master of Science degree in Aeronautical Engineering from the U.S. Naval Postgraduate School in 1964.

MARITAL STATUS: Married to the former Jan Pollom of Topeka, Kansas; her parents, Mr. & Mrs. Harry M. Pollom, reside in Salina, Kansas.

CHILDREN: Jaime D. (daughter), August 21, 1959; Jon P. (son), October 9, 1961

RECREATIONAL INTERESTS: Hobbies include golfing, boating, swimming, fishing, and hunting.

ORGANIZATIONS: Member of Tau Beta Pi, Society of Sigma Xi, and Sigma Nu.

SPECIAL HONORS: Presented the MSC Superior Achievement Award (1970), and winner of eight Air Medals, the Viet Nam Service Medal, and the Navy Commendation Medal with combat distinguishing device.

EXPERIENCE: When notified of his selection to the astronaut program, Evans was on sea duty in the Pacific -- assigned to VF-51 and flying F8 aircraft from the carrier USS TICONDEROGA during a period of seven months in Viet Nam combat operations.

-more-

He was a Combat Flight Instructor (F8 aircraft) with VF-124
from January 1961 to June 1962 and, prior to this assignment,
participated in two WESTPAC aircraft carrier cruises while
a pilot with VF-142. In June 1957, he completed flight
training after receiving his commission as an Ensign through
the Navy ROTC program at the University of Kansas.

Total flight time accrued during his military career is
4,041 hours.

CURRENT ASSIGNMENT: Commander Evans is one of the 19 astronauts
selected by NASA in April 1966. He served as a member of
the astronaut support crews for the Apollo VII and XI flights
and as backup command module for Apollo XIV.

NATIONAL AERONAUTICS AND SPACE ADMINISTRATION

WASHINGTON, D. C. 20546

BIOGRAPHICAL DATA

NAME: Harrison H. Schmitt (PhD)
NASA Astronaut - Apollo 17 lunar module pilot

BIRTHPLACE AND DATE: Born July 3, 1935, in Santa Rita, New Mexico.
His mother, Mrs. Harrison A. Schmitt, resides in Silver City,
New Mexico.

PHYSICAL DESCRIPTION: Black hair; brown eyes; height: 5 feet 9
inches; weight: 165 pounds.

EDUCATION: Graduated from Western High School, Silver City, New
Mexico; received a Bachelor of Science degree in Science
from the California Institute of Technology in 1957; studied
at the University of Oslo in Norway during 1957-58; received
Doctorate in Geology from Harvard University in 1964.

MARITAL STATUS: Single

RECREATIONAL INTERESTS: His hobbies include skiing, hunting, fishing,
carpentry and hiking.

ORGANIZATIONS: Member of the Geological Society of America, the
American Geophysical Union, the American Association for the
Advancement of Science, the American Association of Petro-
leum Geologists, the American Institute of Aeronautics and
Astronautics, and Sigma Xi.

SPECIAL HONORS: Winner of a Fulbright Fellowship (1957-58); a
Kennecott Fellowship in Geology (1958-59); a Harvard Fellowship (1959-60); a Harvard Traveling Fellowship (1960); a
Parker Traveling Fellowship (1961-62); a National Science
Foundation Post-Doctoral Fellowship, Department of Geological Sciences, Harvard University (1963-64); and presented
the MSC Superior Achievement Award (1970).

EXPERIENCE: Schmitt was a teaching fellow at Harvard in 1961; he
assisted in the teaching of a course in ore deposits there.
Prior to his teaching assignment, he did geological work for
the Norwegian Geological Survey in Oslo, Norway, and for the
U.S. Geological Survey in New Mexico and Montanna. He also
worked as a geologist for two summers in southeastern Alaska.

-more-

Before coming to the Manned Spacecraft Center, he served with
the U.S. Geological Survey's Astrogeology Branch at Flagstaff,
Arizona. He was project chief for lunar field geological
methods and participated in photo and telescopic mapping of the
Moon; he was among the USGS astrogeologists instructing NASA
astronauts during their geological field trips. He has
logged more than 1,665 hours flying time.

CURRENT ASSIGNMENT: Dr. Schmitt was selected as a scientist- astronaut by NASA in June 1965. He completed a 53-week course
in flight training at Williams Air Force Base, Arizona,
and, in addition to training for future manned space flights,
has been instrumental in providing Apollo flight crews with
detailed instruction in lunar navigation, geology, and
feature recognition. He has also assisted in the integra-
tion of scientific activities into the Apollo lunar missions
and participated in research activities requiring the conduct
of geologic, petrographic, and stratographic analysis of
samples returned from the Moon by Apollo missions.

Schmitt served as backup lunar module pilot for Apollo XV.

NATIONAL AERONAUTICS AND SPACE ADMINISTRATION

WASHINGTON, D. C. 20546

BIOGRAPHICAL DATA

NAME: John W. Young (Captain, USN)
NASA Astronaut - Apollo 17 backup commander

BIRTHPLACE AND DATE: Born in San Francisco, California, on September 24, 1930. His parents, Mr. and Mrs. William H. Young, reside in Orlando, Florida.

PHYSICAL DESCRIPTION: Brown hair; green eyes; height: 5 feet 9 inches; weight: 165 pounds.

EDUCATION: Graduated from Orlando High School, Orlando, Florida; received a Bachelor of Science degree in Aeronautical Engineering from the Georgia Institute of Technology in 1952; recipient of an Honorary Doctorate of Laws degree from Western State University College of Law in 1969, and an Honorary Doctorate of Applied Science from Florida Technological University in 1970.

MARITAL STATUS: Married to the former Susy Feldman of St. Louis, Missouri.

CHILDREN: Sandy, April 30, 1957; John, January 17, 1959, by a previous marriage.
RECREATIONAL INTERESTS: He plays handball, runs and works out in the full pressure suit to stay in shape.

ORGANIZATIONS: Fellow of the American Astronautical Society, Associate Fellow of the Society of Experimental Test Pilots, and a member of the American Institute of Aeronautics and Astronautics.

SPECIAL HONORS: Awarded the NASA Distinguished Service Medal, two NASA Exceptional Service Medals, the MSC Certificate of Commendation (1970), the Navy Astronaut Wings, the Navy Distinguished Service Medals, and three Navy Distinguished Flying Crosses.

EXPERIENCE: Upon graduation from Georgia Tech, Young entered the U.S. Navy in 1952; he holds the rank of Captain in that service.

-more-

He completed test pilot training at the U.S. Naval Test
Pilot School in 1959, and was then assigned as a test pilot
at the Naval Air Test Center until 1962. Test projects in
which he participated include evaluations of the F8D
"Crusader" and the F4B "Phantom" fighter weapons systems,
and in 1962, he set world time-to-climb records to 3,000
and 25,000 meter altitudes in the Phantom. Prior to his
assignment to NASA, he was maintenance officer of All-Weather-
Fighter Squadron 143 at the Naval Air Station, Miramar,
California.

He has logged more than 6,380 hours flying time, and completed three space flights totaling 267 hours and 42 minutes.

CURRENT ASSIGNMENT: Captain Young was selected as an astronaut
     by NASA in September 1962.

     He served as pilot with command pilot Gus Grissom on the
     first manned Gemini flight -- a 3-orbit mission, launched
     on March 23, 1965, during wihich the crew accomplished the
     first manned spacecraft orbital trajectory modifications
     and lifting reentry, and flight tested all systems in
     Gemini 3.

     After this flight, he was backup pilot for Gemini 6.

     On July 18, 1966, Young occupied the command pilot seat for
     the Gemini 10 mission and, with Michael Collins as pilot,
     effected a successful rendezvous and docking with the Agena
     target vehicle.

     He was then assigned as the backup command module pilot
     for Apollo VII.

     Young was command module pilot for Apollo X, May 18-26, 1969,
     the comprehensive lunar-orbital qualification test of the
     Apollo lunar module. He was accompanied on the 248,000
     nautical mile lunar mission by Thomas P. Stafford (space-
     craft commander) and Eugene A. Cernan (lunar module pilot).

     Captain Young then served as backup spacecraft commander
     for Apollo XIII.

     Young was commander of the Apollo 16 mission to the Descartes
     highlands of the Moon in April 1972.

NATIONAL AERONAUTICS AND SPACE ADMINISTRATION

WASHINGTON. D. C. 20546

BIOGRAPHICAL DATA

NAME: Stuart Allen Roosa (Lieutenant Colonel, USAF)
        NASA Astronaut - Apollo 17 backup command module pilot

BIRTHPLACE AND DATE: Born August 16, 1933, in Durango, Colorado.
        His parents, Mr. and Mrs. Dewey Roosa, now reside in
        Tucson, Arizona.

PHYSICAL DESCRIPTION: Red hair; blue eyes; height: 5 feet 10
        inches; weight: 155 pounds.

EDUCATION: Attended Justice Grade School and Claremore High School
        in Claremore, Oklahoma; studied at Oklahoma State Univer-
        sity of Arizona and was graduated with honors and a Bache-
        lor of Science degree in Aeronautical Engineering from the
        University of Colorado; presented an Honorary Doctorate of
        Letters from the University of St. Thomas (Houston, Texas)
        in 1971.

MARITAL STATUS: His wife is the former Joan C. Barrett of Tupelo,
        Mississippi; and her mother, Mrs. John T. Barrett, resides
        in Sessums, Mississippi.

CHILDREN: Christopher A., June 29, 1959; John D., January 2, 1961;
        Stuart A., Jr., March 12, 1962; Rosemary D., July 23, 1963.

RECREATIONAL INTERESTS: His hobbies are hunting, boating, and
        fishing.

ORGANIZATIONS: Associate Member of the Society of Experimental
        Test Pilots.

SPECIAL HONORS: Presented the NASA Distinguished Service Medal,
        the MSC Superior Achievement Award (1970), the Air Force
        Command Pilot Astronaut Wings, the Air Force Distinguished
        Service Medal, the Arnold Air Society's John F. Kennedy
        Award (1971), and the City of New York Gold Medal in 1971.

EXPERIENCE: Roosa, a Lt. Colonel in the Air Force, has been on
        active duty since 1953. Prior to joining NASA, he was an
        experimental test pilot at Edwards Air Force Base, Cali-
        fornia -- an assignment he held from September 1965 to
        May 1966, following graduation from the Aerospace Research
        Pilots School.

-more-

He was a maintenance flight test pilot at Olmsted Air Force Base, Pennsylvania, from July 1962 to August 1964, flying F-101 aircraft. He served as Cheif of Service Engineering (AFLC) at Tachikawa Air Base for two years following graduation from the University of Colorado under the Air Force Institute of Technology Program. Prior to this tour of duty, he was assigned as a fighter pilot at Langley Air Force Base, Virginia, where he flew the F-84F and F-100 aircraft.

He attended Gunnery School at Del Rio and Luke Air Force Bases and is a graduate of the Aviation Cadet Program at Williams Air Force Base, Arizona, where he received his flight training and commission in the Air Force.

Since 1953, he has acquired 4,797 flying hours.

CURRENT ASSIGNMENT: Lt. Colonel Roosa is one of the 19 astronauts selected by NASA in April 1966. He was a member of the astronaut support crew for the Apollo IX flight.

He completed his first space flight as command module pilot on Apollo XIV, January 31 - February 9, 1971. With him on man's third lunar landing mission were Alan B. Shepard (spacecraft commander) and Edgar D. Mitchell (lunar module pilot). In completing his first space flight, Roosa logged a total of 216 hours and 42 minutes.

He was subsequently designated to serve as backup command module for Apollo XVI.

NATIONAL AERONAUTICS AND SPACE ADMINISTRATION

WASHINGTON, D. C. 20546

BIOGRAPHICAL DATA

NAME: Charles Moss Duke, Jr. (Colonel, USAF)
NASA Astronaut - Apollo 17 backup lunar module pilot.

BIRTHPLACE AND DATE: Born in Charlotte, North Carolina, on October 3, 1935. His parents, Mr. and Mrs. Charles M. Duke, make their home in Lancaster, South Carolina.

PHYSICAL DESCRIPTION: Brown hair; brown eyes; height: 5 feet 11 1/2 inches; weight: 155 pounds.

EDUCATION: Attended Lancaster High School in Lancaster, South Carolina, and was graduated valedictorian from the Admiral Farragut Academy in St. Petersburg, Florida; received a Bachelor of Science degree in Naval Sciences from the U.S. Navel Academy in 1957 and a Master of Science degree in Aeronautics from the Massachusetts Institute of Technology in 1964.

MARITAL STATUS: Married to the former Dorothy Meade Claiborne of Atlanta, Georgia; her parents are Dr. and Mrs. T. Sterling Claiborne of Atlanta.

CHILDREN: Charles M., March 8, 1965; Thomas C., May 1, 1967.

RECREATIONAL INTERESTS: Hobbies include hunting, fishing, reading, and playing golf.

ORGANIZATIONS: Member of the Air Force Association, the Society of Experimental Test Pilots, the Rotary Club, the American Legion, and the American Fighter Pilots Association.

SPECIAL HONORS: Awarded the MSC Certificate of Commendation (1970)

EXPERIENCE: When notified of his selection as an astronaut, Duke was at the Air Force Aerospace Research Pilot School as an instructor teaching control systems and flying in the F-104, F-101, and T-33 aircraft. He was graduated from the Aerospace Research Pilot School in September 1965 and stayed on there as an instructor.

-more-

He is an Air Force Colonel and was commissioned in 1957 upon graduation from the Naval Academy. Upon entering the Air Force, he went to Spence Air Base, Georgia, for primary flight training and then to Webb Air Force Base, Texas, for basic flying training, where in 1958 he became a distinguished graduate. He was again a distinguished graduate at Moody Air Force Base, Georgia, where he completed advanced training in F-86L aircraft. Upon completion of this training he was assigned to the 526th Fighter Interceptor Squadron at Ramstein Air Base, Germany, where he served three years as a fighter interceptor pilot.

He has logged 3,862 hours flying time.

CURRENT ASSIGNMENT: Colonel Duke is one of the 19 astronauts selected by NASA in April 1966. He served as a member of the astronaut support crew for the Apollo X flight and as backup lunar module pilot for the Apollo XIII flight.

He served as lunar module pilot for the Apollo 16 mission.

SPACEFLIGHT TRACKING AND DATA SUPPORT NETWORK

NASA's worldwide Spaceflight Tracking and Data Network (STDN) will provide communication with the Apollo astronauts, their launch vehicle and spacecraft. It will also maintain the communications link between Earth and the Apollo experiments left on the lunar surface by earlier Apollo crews.

The STDN is linked together by the NASA Communication Network (NASCOM) which provides for all information and data flow.

In support of Apollo 17, the STDN will employ 11 ground tracking stations equipped with 9.1-meter (30-foot) and 25.9 m (85-ft) antennas, and instrumented tracking ship, and four instrumented aircraft. This portion of the STDN was known formerly as the Manned Space Flight Network. For Apollo 17, the network will be augmented by the 64-m (210-ft.) antenna system at Goldstone, Calif. (a unit of NASA's Deep Space Netwrok), and if required the 64-m (210-ft.) radio antenna of the National Radio Astronomy Observatory at Parkes, Australia.

The STDN is maintained and operated by the NASA Goddard Space Flight Center, Greenbelt, Md., under the direction of NASA's Office of Tracking and Data Acquisition. Goddard will become an emergency control center if the Houston Mission Control Center is impaired for an extended time.

NASA Communications Network (NASCOM). The tracking network is linked together by the NASA Communications Network. All information flows to and from Mission Control Center, (MCC), Houston, and the Apollo spacecraft over this communications system.

The NASCOM consists of more than 3.2 million circuit kilometers (1.7 million nautical miles), using satellites, submarine cables, land lines, microwave systems, and high frequency radio facilities. NASCOM control center is located at Goddard. Regional communication switching centers are in Madrid; Canberra, Australia; Honolulu; and Guam.

Intelsat communications satellites will be used for Apollo 17. One satellite over the Atlantic will link Goddard with Ascension Island and the Vanguard tracking ship. Another Atlantic satellite will provide a direct link between Madrid and Goddard for TV signals received from the spacecraft. One satellite positioned over the mid-Pacific will link Carnarvon, Australia; Canberra, Guam and Hawaii with Goddard through the Jamesburg, California ground station. An alternate route of communications between Spain and Australia is available through another Intelsat satellite positioned over the Indian Ocean if required.

Mission Operations: Prelaunch tests, liftoff, and Earth orbital flight of the Apollo 17 are supported by the Apollo subnet station at Merritt Island, Fla., 6.4 km (3.5 nm) from the launch pad.

During the critical period of launch and insertion of the Apollo 17 into Earth orbit, the USNS Vanguard provides tracking, telemetry, and communications functions. This single sea-going station of the Apollo subnet will be stationed about 1,610 km (870 nm) southeast of Bermuda.

When the Apollo 17 conducts the translunar injection (TLI) Earth orbit for the Moon, two Apollo range instrumentation aircraft (ARIA) will record telemetry data from Apollo and relay voice communications between the astronauts and the MCC at Houston. These aircraft will be airborne between South America and the west coast of Africa. ARIA 1 will cover TLI ignition and ARIA 2 will monitor TLI burn completion.

Approximately 1 hour after the spacecraft has been injected into a translunar trajectory, three prime MSFN stations will take over tracking and communication with Apollo. These stations are equipped with 25.9 m (85-ft.) antennas.

Each of the prime stations, located at Goldstone, Madrid, and Honeysuckle, Australia is equipped with dual systems for tracking the command module in lunar orbit and the lunar module in separate flight paths or at rest on the Moon.

For reentry, two ARIA (Apollo Range Instrumented Aircraft) will be deployed to the landing area to relay communications between Apollo and Mission Control at Houston. These aircraft also will provide position information on the Apollo after the blackout phase of reentry has passed.

An applications technology satellite (ATS) terminal has been placed aboard the recovery ship USS Ticonderoga to relay command control communications of the recovery forces, via NASA's ATS satellite. Communications will be relayed from the deck-mounted terminal to the NASA tracking stations at Mojave, Calif. and Rosman, N.C., through Goddard to the recovery control centers located in Hawaii and Houston.

Prior to recovery, the astronauts aeromedical records are transmitted via the ATS satellite to the recovery ship for comparison with the physical data obtained in the postflight examination performed aboard the recovery ship.

Television Transmissions: Television from the Apollo spacecraft during the journey to and from the Moon and on the lunar surface will be received by the three prime stations, augmented by the 64-m (210-ft.) antennas at Goldstone and Parkes. The color TV signal must be converted at MSC, Houston. A black and white version of the color signal can be released locally from the stations in Spain and Australia.

While the camera is mounted on the lunar roving vehicle (LRV), the TV signals will be transmitted directly to tracking stations as the astronauts explore the Moon.

Once the LRV has been parked near the lunar module, its batteries will have about 80 hours of operating life. This will allow ground controllers to position the camera for viewing the lunar module liftoff, post liftoff geology, and other scenes.

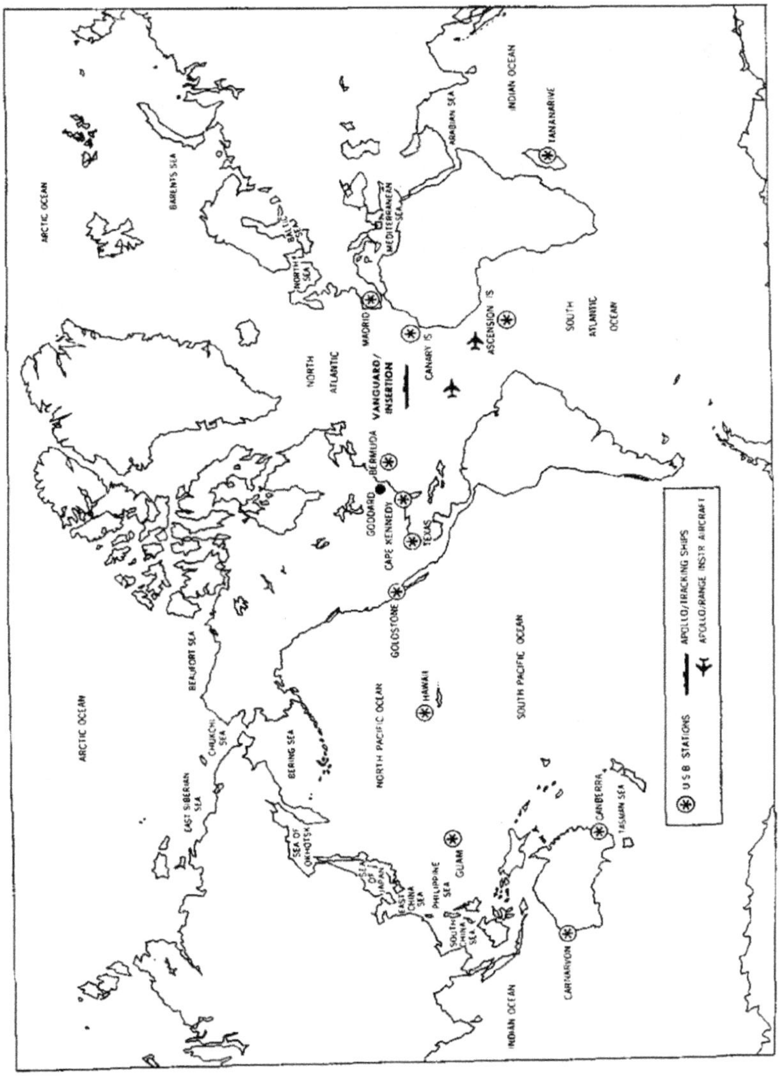
ARCTIC OCEAN
BARENTS SEA
NORTH SEA
BALTIC SEA
MEDITERRANEAN SEA
ARABIAN SEA
INDIAN OCEAN
TANANARIVE
MADRID
CANARY IS
ASCENSION IS
NORTH ATLANTIC
SOUTH ATLANTIC OCEAN
VANGUARD/INSERTION
BERMUDA
GODDARD
CAPE KENNEDY
TEXAS
GOLDSTONE
HAWAII
NORTH PACIFIC OCEAN
SOUTH PACIFIC OCEAN
BEAUFORT SEA
CHUKCHI SEA
BERING SEA
ARCTIC OCEAN
EAST SIBERIAN SEA
SEA OF OKHOTSK
SEA OF JAPAN
EAST CHINA SEA
PHILIPPINE SEA
GUAM
SOUTH CHINA SEA
CARNARVON
CANBERRA
TASMAN SEA
INDIAN OCEAN
USB STATIONS
APOLLO/TRACKING SHIPS
APOLLO/RANGE INSTR AIRCRAFT

ENVIRONMENTAL IMPACT OF APOLLO/SATURN V MISSION

Studies of NASA space mission operations have concluded that Apollo does not significantly effect the human environment in the areas of air, water, noise or nuclear radiation.

During the launch of the Apollo/Saturn V space vehicle, products exhausted from Saturn first stage engines in all cases are within an ample margin of safety. At lower altitudes, where toxicity is of concern, the carbon monoxide is oxidized to carbon dioxide upon exposure at its high temperature to the surrounding air. The quantities released are two or more orders of magnitude below the recognized levels for concern in regard to significant modification of the environment. The second and third stage main propulsion systems generate only water and a small amount of hydrogen. Solid propellant ullage and retro rocket products are released and rapidly dispersed in the upper atmosphere at altitudes above 70 kilometers (43.5 miles). This material will effectively never reach sea level and, consequently, poses no toxicity hazard.

Should an abort after launch be necessary, some RP-1 fuel (kerosene) could reach the ocean. However, toxicity of RP-1 is slight and impact on marine life and waterfowl are considered negligible due to its dispersive characteristics. Calculations of dumping an aborted S-IC stage into the ocean showed that spreading and evaporating of the fuel occurred in one to four hours.

There are only two times during a nominal Apollo mission when above normal overall sound pressure levels are encountered. These two times are during vehicle boost from the launch pad and the sonic boom experienced when the spacecraft enters the Earth's atmosphere. Sonic boom is not a significant nuisance since it occurs over the mid-Pacific Ocean.

NASA and the Department of Defense have made a comprehensive study of noise levels and other hazards to be encountered for launching vehicles of the Saturn V magnitude. For uncontrolled areas the overall sound pressure levels are well below those which cause damage or discomfort. Saturn launches have had no deleterious effects on wildlife which has actually increased in the NASA-protected areas of Merritt Island.

-more-

A source of potential radiation hazard but highly unlikely, is the fuel capsule of the radioisotope thermoelectric
generator supplied by the Atomic Energy Commission which
provides electric power for Apollo lunar surface experiments.
The fuel cask is designed to contain the nuclear fuel during
normal operations and in the event of aborts so that the
possibility of radiation contamination is negligible. Extensive
safety analyses and tests have been conducted which demonstrated
that the fuel would be safely contained under almost all credible accident conditions.

PROGRAM MANAGEMENT

The Apollo Program is the responsibility of the Office of Manned Space Flight (OMSF), National Aeronautics and Space Administration, Washington, D. C. Dale D. Myers is Associate Administrator for Manned Space Flight.

NASA Manned Spacecraft Center (MSC), Houston, is responsible for development of the Apollo spacecraft, flight crew training, and flight control. Dr. Christopher C. Kraft, Jr. is Center Director.

NASA Marshall Space Flight Center (MSFC), Huntsville, Ala., is responsible for development of the Saturn launch vehicles. Dr. Eberhard F. M. Rees is Center Director.

NASA John F. Kennedy Space Center (KSC), Fla., is responsible for Apollo/Saturn launch operations. Dr. Kurt H. Debus is Center Director.

The NASA Office of Tracking and Data Acquisition (OTDA) directs the program of tracking and data flow on Apollo. Gerald M. Truszynski is Associate Administrator for Tracking and Data Acquisition.

NASA Goddard Space Flight Center (GSFC), Greenbelt, Md., manages the Manned Space Flight Network and Communications Network. Dr. John F. Clark is Center Director.

The Department of Defense is supporting NASA during launch, tracking, and recovery operations. The Air Force Eastern Test Range is responsible for range activities during launch and down-range tracking. Recovery operations include the use of recovery ships and Navy and Air Force aircraft.

APOLLO/SATURN OFFICIALS

NASA Headquarters

Dr. Rocco A. Petrone	Apollo Program Director, OMSF
Chester M. Lee (Capt., USN, Ret.)	Apollo Mission Director, OMSF
John K. Holcomb (Capt., USN, Ret.)	Director of Apollo Operations, OMSF
William T. O'Bryant (Capt., USN, Ret.)	Director of Apollo Lunar Exploration, OMSF
Charles A. Berry, M.D.	Director for Life Sciences

Kennedy Space Center

Miles J. Ross	Deputy Center Director
Walter J. Kapryan	Director of Launch Operations
Peter A. Minderman	Director of Technical Support
Robert C. Hock	Apollo/Skylab Program Manager
Dr. Robert H. Gray	Deputy Director, Launch Operations
Dr. Hans F. Gruene	Director, Launch Vehicle Operations
John J. Williams	Director, Spacecraft Operations
Paul C. Donnelly	Associate Director Launch Operations
Isom A. Rigell	Deputy Director Launch Vehicle Operations

Manned Spacecraft Center

Sigurd A. Sjoberg	Deputy Center Director
Howard W. Tindall	Director, Flight Operations
Owen G. Morris	Manager, Apollo Spacecraft Program
Donald K. Slayton	Director, Flight Crew Operations
Pete Frank	Flight Director
Neil Hutchinson	Flight Director
Gerald D. Griffin	Flight Director
Eugene F. Kranz	Flight Director
Charles Lewis	Flight Director
Richard S. Johnston	Director, Life Sciences

- more -

Marshall Space Flight Center

Dr. William R. Lucas	Deputy Center Director, Technical
Richard W. Cook	Deputy Center Director, Management
Richard G. Smith	Manager, Saturn Program Office (SPO)
John C. Rains	Manager, S-IC Stage Project, SPO
William F. LaHatte	Manager, S-II, S-IVB Stage Projects SPO
James B. Bramlet	Manager, Instrument Unit, GSE Projects SPO
James M. Sisson	Manager, LRV Project, SPO
T.P. Smith	Manager, Engines Project, SPO
Herman F. Kurtz	Manager, Mission Operations Office

Goddard Space Flight Center

Tecwyn Roberts	Director, Networks
William P. Varson	Chief, Network Computing & Analysis Division
Walter Lafleur	Chief, Network Operations Division
Robert Owen	Chief, Network Engineering Division
L.R. Stelter	Chief, NASA Communications Division

Department of Defense

Maj. Gen. David M. Jones, USAF	DOD Manager for Manned Space Flight Support Operations
Col. Alan R. Vette, USAF	Deputy DOD Manager for Manned Space Flight Support Operations, and Director, DOD Manned Space Flight Support Office
Rear Adm. J.L. Bulls USN	Commander, Task Force 130, Pacific Recovery Area
Rear Adm. Roy G. Anderson, USN	Commander Task Force 140, Atlantic Recovery Area
Capt. Norman K. Green, USN	Commanding Officer, USS Ticonderoga, CVS-14 Primary Recovery Ship
Brig. Gen. Frank K. Everest, Jr. USAF	Commander Aerospace Rescue and Recovery Service

CONVERSION TABLE

	Multiply	By	To Obtain
Distance:	inches	2.54	centimeters
	feet	0.3048	meters
	meters	3.281	feet
	kilometers	3281	feet
	kilometers	0.6214	statute miles
	statute miles	1.609	kilometers
	nautical miles	1.852	kilometers
	nautical miles	1.1508	statute miles
	statute miles	0.8689	nautical miles
	statute miles	1760	yards
Velocity:	feet/sec	0.3048	meters/sec
	meters/sec	3.281	feet/sec
	meters/sec	2.237	statute mph
	feet/sec	0.6818	statute miles/hr
	feet/sec	0.5925	nautical miles/hr
	statute miles/hr	1.609	km/hr
	nautical miles/hr (knots)	1.852	km/hr
	km/hr	0.6214	statute miles/hr
Liquid measure, weight:	gallons	3.785	liters
	liters	0.2642	gallons
	pounds	0.4536	kilograms
	kilograms	2.205	pounds
	metric ton	1000	kilograms
	short ton	907.2	kilograms
Volume:	cubic feet	0.02832	cubic meters
Pressure:	pounds/sq. inch	70.31	grams/sq.cm
Thrust:	pounds	4.448	newtons
	newtons	0.225	pounds
Temperature:	Centigrade	1.8· add 32	Fahrenheit

-end-